Military Engineering

Edited by George Dekoulis

Published in London, United Kingdom

IntechOpen

Supporting open minds since 2005

Military Engineering
http://dx.doi.org/10.5772/intechopen.79057
Edited by George Dekoulis

Contributors
Jian Chen, Yu Han, Yuan Ren, Walter Fuertes, César Villacís, Fabián Romero, Luis Escobar, Santiago Chamorro, Kanchan Biswas, George Melillos, Kyriacos Themistocleous, Athos Agapiou, Silas Michaelides, Diofantos Hadjimitsis, Josep Lluis i Ginovart, Michael Naor

First published in London, United Kingdom, 2020 by IntechOpen
IntechOpen is the global imprint of INTECHOPEN LIMITED, registered in England and Wales, registration number: 11086078, 7th floor, 10 Lower Thames Street, London, EC3R 6AF, United Kingdom
Printed in Croatia

British Library Cataloguing-in-Publication Data
A catalogue record for this book is available from the British Library

Additional hard and PDF copies can be obtained from orders@intechopen.com

Military Engineering
Edited by George Dekoulis
p. cm.
Print ISBN 978-1-78923-953-9
Online ISBN 978-1-78923-954-6
eBook (PDF) ISBN 978-1-83880-316-2

For EU product safety concerns:
IN TECH d.o.o., Prolaz Marije Krucifikse Kozulić 3, 51000 Rijeka, Croatia, info@intechopen.com or visit our website at intechopen.com.

Meet the editor

Prof. George Dekoulis received his PhD in space engineering and communications from Lancaster University, UK, in 2007. He was awarded a 1st Class BEng (Hons) degree in communications engineering from De Montfort University, UK, in 2001. He has received several awards from STFC, UK and EPSRC, UK, and the "IET Hudswell International Research Scholarship". He is currently a professor at the Aerospace Engineering Institute (AEI), Cyprus. He is founder of the IEEE Aerospace and Electronic Systems Society (AESS) – Cyprus and was the General Chair of IEEE Aerospace Engineering Innovations 2019 (IEEE AEI 2019) Symposium, 20-23 April 2019, Limassol, Cyprus. He has previously worked as a professor in aerospace engineering at various departments, such as space and planetary physics, aeronautical and space engineering, professional flight, robotics/mechatronics and mechanical engineering, computer science and engineering and electrical and electronics engineering. His research is focused on the design of reconfigurable aerospace engineering systems.

Contents

Preface

This book is a collection of reviewed and relevant research chapters, concerning the developments within the military engineering field of study. The book includes scholarly contributions by various authors and has been edited by a group of experts in physical sciences, engineering and technology. Each contribution comes as a separate chapter, complete in itself but directly related to the book's topics and objectives.

The book includes chapters dealing with the topics: Military Aviation Principles, Scientific Knowledge of Spanish Military Engineers in the Seventeenth Century, Healthcare Military Logistics at Disaster Regions around the World: Insights from Ten Field Hospital Missions over Three Decades, A New Real-Time Flight Simulator for Military Training Using Mechatronics and Cyber-Physical System Methods, Detecting Underground Military Structures Using Field Spectroscopy, and Robust Guidance Algorithm against Hypersonic Targets.

The target audience comprises scholars and specialists in the field.

George Dekoulis
Aerospace Engineering Institute,
Cyprus

Chapter 1

Military Aviation Principles

Kanchan Biswas

Abstract

Military all over the world uses military aircraft in both offensive and defensive purposes. In offensive role, these aircraft are used in destroying enemy's vital installations, air strips, ordnance depots and supplies. In defensive role, it provides close air support to land-based army and also deters the threats of enemy air strike. In naval warfare, military aircraft plays a significant role to detect and neutralize submarines and warships to keep the seacoast free from enemy attack. Military aircraft also provides logistic supply to forward bases, conducting airlift (cargo and troops), and participates in rescue operations during national disaster. Military aviation includes both transport and warcraft and consisting of fixed wing aircraft, rotary-wing aircraft (RWA) and unmanned aerial vehicle (UAV). From the early days of world war, it has been realized that air power supremacy is vital for winning a war as well as maintaining the sovereignty of any country. This chapter discusses basic flight mechanics, types and roles of aircraft, safety considerations and design and certification procedures.

Keywords: military, aviation, combat, aircraft, aerodynamics, helicopter, UAV

1. Introduction

It was realized that aviation had a great potential in transporting goods as well as passengers in large distances in minimum possible time. The military also realized the advantages of having an offensive and defensive air power during the war and peace time. Today air power has become the essence of military supremacy of any country for maintaining country sovereignty during peace time and offensive attack capability to win war by destroying enemy vital installations, deterring troop transfer and military supplies.

The aircraft can be broadly categorized as lighter-than-air aircraft (balloons and airships) which generate lift due the buoyancy forces and heavier-than-air aircraft (aircraft, helicopters and UAV).

2. Principles of flight

An aircraft is a complex machine using the application of multidisciplinary engineering sciences. The major engineering groups can be indicated as follows:

a. Aerodynamics and flight mechanics determining aircraft shape, configuration and control law

b. Airframe structure (fuselage, wings, vertical and horizontal tail planes and control surfaces)

c. Mechanical system (hydraulics, pneumatics, landing gear, fuel and flight control systems)

d. Engine/power plant system

e. Electrical system (power generations, distribution and emergency power)

f. Avionic (communication, navigation, weapon aiming, displays and warnings and utility management system) and instrument system

g. Environmental systems (air-conditioning, life support system and cabin pressurization systems)

h. Armament system (for military aircraft)

i. Air egress (ejection) system

j. Software (embedded as well as operational software)

2.1 Atmosphere

For the aerodynamic study, air is considered an ideal gas which follows gas laws. Therefore, the variation of air properties with respect to altitude is important. The International Standard Atmosphere (ISA) is used for comparison of performances of aircraft designed by different countries. The ISA is defined as:

Altitude (H): sea level (0 m)
Temperature (T): 288.15°K
Pressure (P): 1.01325 MPa (14.7 psi)
Density (ρ): 1.225 kg/m^3

The atmosphere is divided in two layers. The lower layer is called 'troposphere' where temperature decreases linearly with altitude (6.5°C/km altitude rise, known as lapse rate). Air pressure decreases with altitude as shown in **Figure 1**. Air density can be estimated from gas equation ($\rho = P/RT$, where R is universal gas constant). **Figure 1** also shows the different types of clouds in the atmosphere [1–5].

The upper layer of the atmosphere is called 'stratosphere' where T remains constant at −56°C (ISA condition). In troposphere, with increase in altitude, both pressure and temperature decrease reducing the density, thus lowering engine mass flow reducing thrust, lift and drag. These values optimize at 10–12 km which is known as cruise altitude for jet aircraft. Due to decrease in mass flow, jet engines cannot operate at very high altitude.

2.2 Aerodynamics and flight mechanics

Movement of air over an aircraft generates aerodynamic forces and moments. Due to change in the air properties, the aerodynamic forces and moments also vary with altitude. Flight dynamics looks at these aerodynamic forces and includes thrust and gravity forces to study aircraft motion. Further the aerodynamic forces generated due to deflections of control surfaces are added as applied control forces to study the dynamics of the flight path including stability and controllability of the aircraft [1, 5].

Airflow over a body obeys three basic aerodynamic equations: these are conservation of mass, conservation of momentum and conservation of energy. Solving these equations, we obtain velocity distribution over the aircraft surfaces from which forces and moments can be estimated.

Airflow over an airfoil (cross section of wing) is shown in **Figure 2**. Due to the airfoil camber, air particles traveling over the upper surface have to cover longer distance than the air flowing on the lower surface. In order to comply with the law of conservation of mass, air particle on the upper surface speeds up to cover longer distance (due to camber) than the airflow over the lower surface. According to the law of conservation of momentum, the increase in speed is compensated by the decrease in pressure. This creates differences in pressures with lower surface air pressure being more than that of the upper surface. This differential pressure gives rise to upward force. The vertical component of this upward force is the wing lift, and axial component is the drag. Further the resultant of the air pressure of the upper surface and lower surface does not pass through the same point, which creates a turning moment known as pitching moment. An airfoil of unit thickness will produce lift, drag and pitching moment coefficients. These are dimensionless quantities and are represented by C_L, C_D, and C_M.

Aerodynamics of supersonic flow is, however, different. A supersonic flow over an airfoil at angle of attack (AOA) is shown in **Figure 3**. On the upper surface, flow

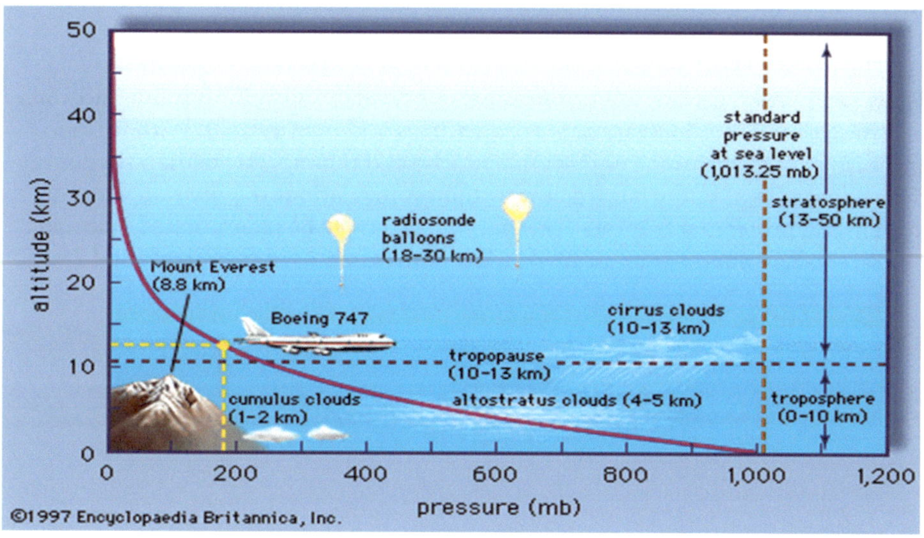

Figure 1.
Variation of pressure of atmospheric pressure with altitude.

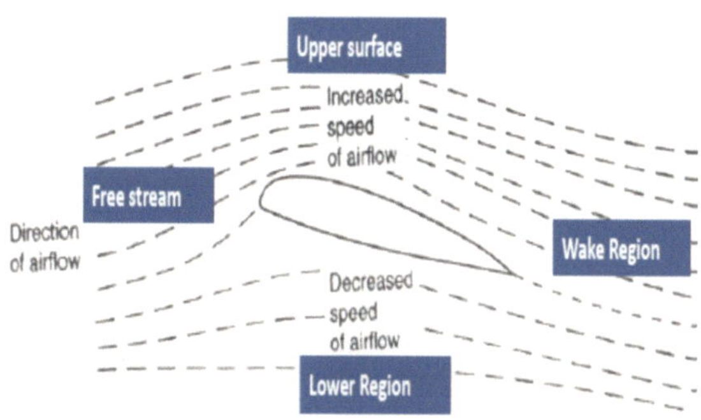

Figure 2.
Subsonic flow over [2].

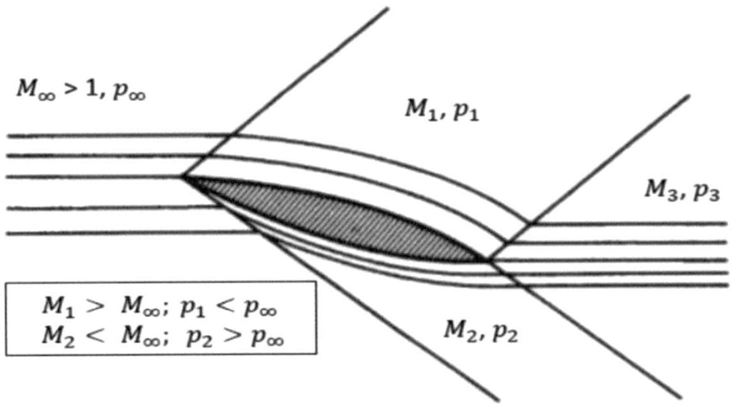

Figure 3.
Supersonic flow [2].

expands having higher flow area, and flow on the lower surface gets compressed due to lower area of flow. Expansion of flow is associated with increase in Mach number and decrease in static pressure. It may be appreciated that at zero AOA, there will be no difference in upper and lower surface flow. As the supersonic airfoil generates lifts only due to AOA, supersonic airfoils are symmetrical airfoils.

The lift produced increases with increase in AOA as shown in **Figure 4**. However, beyond certain AOA, flow separates from the airfoil, lift suddenly drops, and drag rises due to increase of wake area. This is known as 'stall'. With further increase of AOA, the wake region will increase and thus aggravate the situation. This being a flight safety hazard, airworthiness regulations require stall warning system and protection from stall recovery procedure to be incorporated. Subsonic aircraft stalls around 14–18° of AOA, while supersonic aircraft stalls around 24–28°

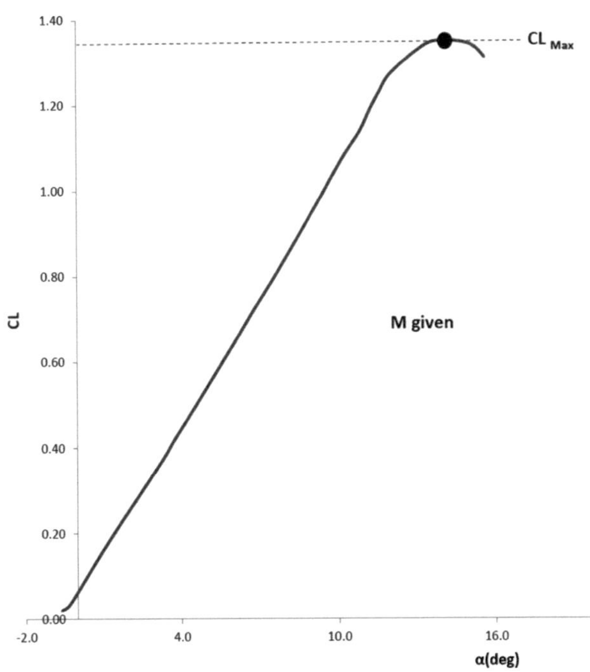

Figure 4.
Stall phenomenon [4].

AOA. In supersonic flow, lift curve slope is very flat, and realization of stall is rather difficult. In view of this in supersonic aircraft, in addition to an AOA indicator, audio warning with stick shaker is installed.

An aircraft in flight has six degrees of freedom (DOF); these are three translations (motion in forward, lateral and vertical directions) and three angular motions, viz. roll (rotation about longitudinal axis), pitch (rotation around lateral axis) and yaw (rotation about vertical axis). An aircraft is steered to the desired direction by operating the respective control surface. This is shown in **Figure 5**.

The lift, drag and pitching moment produced by the wings can be written as $L = \frac{1}{2}\rho V^2 S C_L$; $D = \frac{1}{2}\rho V^2 S C_D$; and $M = \frac{1}{2}\rho V^2 S C C_M$, where S is the total wetted wing area and C is the mean aerodynamic chord length. The force and moments acting on an aircraft in x-z plane during flight are shown in **Figure 6**. The lift, drag and pitching moments act at the aerodynamic centre of the aircraft. The forces and moments are in equilibrium. If the equilibrium is disturbed, the aircraft will move from its flight path. For example, if the lift is increased more than weight, the aircraft will float up, and if the thrust produced is more than the drag, the aircraft will accelerate. The reverses are also true. Similarly, if any control surface is deflected, ΣM about the axis will be disturbed, and the aircraft will rotate about that axis. The degrees of turn of flight path achieved per degree of deflection of control surfaces are the measure of controllability of the aircraft.

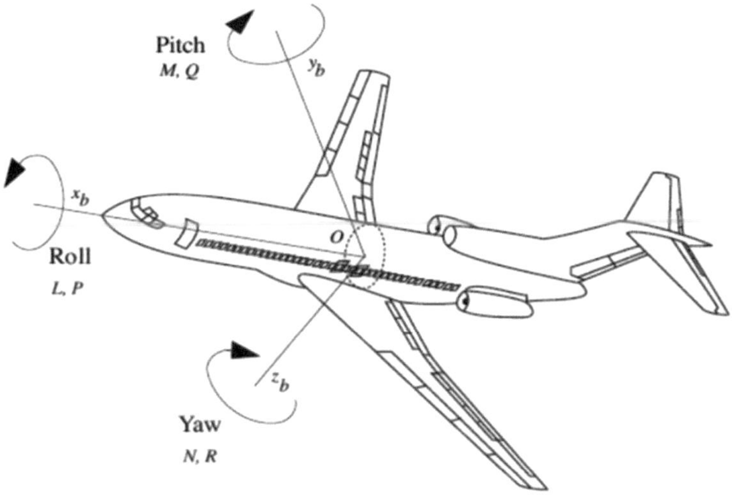

Figure 5.
6 DOF of air vehicle [4].

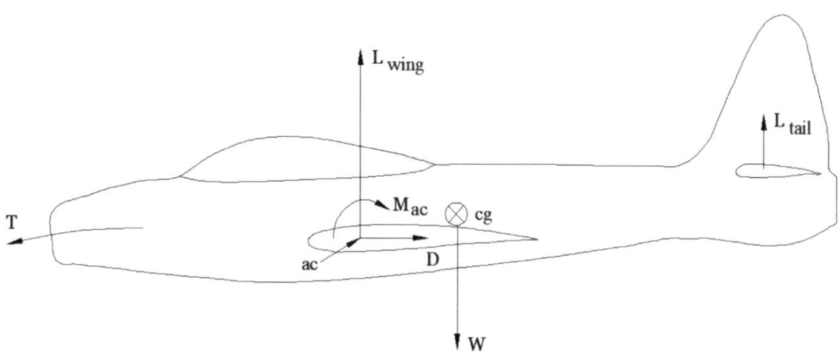

Figure 6.
The force and moment system working on an aircraft in flight [2].

The ease with which an aircraft can be operated is judged by the handling qualities (HQ) of an aircraft. The HQ is directly related to the aircraft stability and controllability. An aircraft is said to be statically stable if an aircraft while flying in a steady path, if unintentionally disturbed by some external forces like gust or any other reason, the aerodynamic forces and moments so created due to the disturbances bring the aircraft back to its original stable condition. This property of the aircraft is termed as stability. However, higher the stability, higher will be the demand for control power to steer the aircraft and lower will be the controllability of the aircraft. Thus, a compromise is made, and the stability requirements are specified in design regulations formulated by the airworthiness authorities. The advantages of lower stability have brought the concept of 'relaxed static stability' or 'statically unstable' aircraft. The high-performance military fighters like F-16, F-17 and F-18 are statically unstable in order to obtain dramatic increase in manoeuverability. The vehicle is kept under control by 'fly-by-wire' (FBW) control system. In FBW, accelerometers and rate gyros are mounted in each axis which senses the aircraft position and attitude, and a FBW computer continuously monitors these data and commands the control actuators to move the control surfaces to keep the aircraft under control. In this approach, very high manoeuverability advantages can be realized without heavily taxing the pilot. Even in the transport aircraft, this leads to smaller tailplane resulting in lesser weight and drag. The light combat aircraft of India, Mirage-2000 and SU-30, belong to this category of unstable aircraft [1–5].

2.3 Flight Mach number

At low flight speeds, compressibility of air may be neglected (ρ assumed constant), but as the flight speed increases, air gets compressed, and change in ρ cannot be neglected. Above $M > 0.3$ (M = Mach number; defined as the ratio of aircraft speed and speed of sound, named after Austrian Physicist Ernst Mach), compressibility effect cannot be neglected. The solution of a three-dimensional flow with air viscosity terms included at higher speed thus becomes complex.

Depending on the flow M, the aerodynamic studies are classified as:

a. *Subsonic flow*—the flight M No. is 0–0.8.

b. *Transonic flow*—the flight M No. is 0.8–1.2. In this flow, the local M No. over some parts of the airfoil is supersonic, while it is subsonic in some other parts. Transonic flight regime is associated with very high drag and instability due to formation of shock and shock oscillation. No aircraft therefore does sustain flight in transonic regime.

c. *Supersonic flow*—the flight $M > 1.2$. In supersonic flow the normal shock formed at $M = 1$ becomes bow shock forming a conical region over the aircraft. The half angle of the cone is $\sin^{-1}\frac{1}{M}$. Due to formation of shock, supersonic drag is 50–80% higher than the subsonic drags. The drag arises due to the thickness effect, and a hypothetical zero-thickness flat-plate airfoil in supersonic flow will produce no drag. However, practical aircraft wings required adequate thickness for storage of fuel in wing tanks and to provide adequate structural stiffness.

d. *Hypersonic flow*—the flight M No. >7. When an airflow is brought to rest (like in the leading edges of wings, fuselages, tail planes, etc.), the kinetic energy of air is converted to thermal energy and temperature rises; this is called stagnation temperature (T_s). T_s can be estimated from the equation $T_s = T_0[1 + 0.2M^2]$. For flow with $M = 7$ and above, the temperature rise may be so high that the air may be ionized and ideal gas equations may no more be applicable, and we may use equations from gas dynamics.

3. Military aircraft

Any aircraft that is operated by a legal or insurrectionary armed force may be called military aircraft. Military uses aircraft for both combat and noncombat applications.

3.1 Combat aircraft

Combat aircraft are designed and developed for use by military to destroy enemy assets using on-board armaments/stores. Military aircraft and their applications include the following [1, 3, 5].

3.1.1 Fighters (air superiority, interceptor and fighter)

Fighters are meant to engage in air-to-air combat with enemy aircraft and outclass them. They are therefore light and have high speeds and manoeuverability. Fighters are used for both offensive and defensive roles. Interceptor is intended to be light and agile and has high acceleration and rate of climb to intercept an enemy aircraft spotted by the ground surveillance radar and engage in dogfight. The main weapons these aircraft carry are air-to-air combat missile and air gun. Many fighters have a secondary role of ground attack capabilities where it carries bombs and air-to-surface missiles (ASM). This is how the concept of multirole combat aircraft (MRCA) has evolved.

A fighter's main purpose is to establish and maintain 'air superiority' which means it denies the air power of opposing air forces for effective interference. Since the early days of aerial combat, armed forces have constantly invested to develop technologically superior fighters and attain air supremacy over the adversaries. Substantial proportion of the defence budgets of modern armed forces is spent for these purposes. Some of the modern fighters include General Dynamics F-16 Fighting Falcon, Dassault Rafale, Dassault Mirage 2000, Russian Su-30MKI, Mikoyan MiG-29, Saab JAS 39 Gripen, etc.

3.1.2 Bombers (bombers, strategic bombers, tactical bomber and interdictor)

A bomber is a combat aircraft designed to attack ground and naval targets by dropping air-to-ground weapons like bombs (dumb and smart bombs); firing ASM, torpedoes and bullets; or deploying air-launched cruise missiles. Heavy bombers (known as strategic bombers) armed with powerful conventional or nuclear weapons are used for long-range bombing missions against strategic targets such as supply bases, bridges, factories, shipyards and cities and thus cripple the enemy infrastructure and capability to continue war or stage new attacks. B-2 Spirit, B-52, Tupolev TU-95, etc. are some of the present days' strategic bombers.

Tactical bombers are smaller bombers with shorter range and weapon capability and used for battlefield tactical operations like countering enemy military activity and military troop transport/supplies and supporting offensive operations.

An interdictor is an attack aircraft designed to interrupt enemy supply operation by aerial bombing. A deep penetration aircraft is a version of interdictor having longer range and capabilities. The main purpose of these aircraft is to prevent or cause delays to enemy forces and supplies reaching the battlefront. Russian MiG-23BN/MiG-27, Dassault-Breguet Mirage 2000D and Panavia Tornado are some of the present-day bombers.

3.1.3 Multirole combat aircraft (multirole, fighter bomber, strike fighter)

The MRCA roles may include air-to-air combat, bombing operation, aerial photo reconnaissance, etc. The main motivation for developing multirole aircraft is cost reduction in using a common airframe. Multirole means an aircraft with major roles, a primary air-to-air combat role and a secondary role like air-to-surface attack. It may be appreciated that an aircraft optimized for a particular role may not fulfil some other role efficiently; however, they may be gainfully used for secondary roles. For example, air superiority fighters designed for higher manoeuverability over the contemporary aircraft are also capable of ground attack. Similarly an interceptor having high rate of climb and acceleration and equipped with close combat missiles and air guns can also be used to chase enemy aircraft and neutralize them in air. Aircraft roles can be changed by changing on-board armament stores/role equipment. Some aircraft are retro modified for additional role. The F-14 was envisioned originally for air superiority and fleet interception defence with some variants later receiving secondary ground attack capability.

The Euro fighter Typhoon and Dassault Rafal are classified as multirole fighters. Euro fighter Typhoon was however, originally designed as an air superiority fighter.

3.1.4 Electronic warfare (EW) aircraft

An EW aircraft is a military aircraft equipped with EW system meant to degrade the effectiveness of enemy radar and radio systems by using radar jamming deception methods. EW is a technology to detect, identify the frequency spectrum and locate the source of electromagnetic energy and use of self-protective jammer to deny the opponent access to the EM spectrum. EW can be deployed from land, sea or air. Various types of self-protective jammers like noise/frequency jamming and counter-measure dispensing systems (CMDS) are used. This gives the aircraft stealth capability to deceive the enemy radars. F-35 has very advanced EW capability which enables it to reach well-defended targets and suppress enemy radars that threaten the F-35.

3.1.5 Maritime patrol aircraft

A maritime patrol aircraft is designed to operate for long duration over coastline/water territory in order to detect, identify enemy ships and submarines and destroy them using air-to-surface weapon, torpedoes and underwater mines. These aircraft are equipped with various sensors including sonar and radars and are also used in maritime search and rescue operations. Jaguar maritime patrol aircraft is one such aircraft that is in use for a long time.

3.2 Fighter aircraft generations

Aircraft is designed for an economic life of 20 years from the consideration of obsolescence. However, as the cost of procurement of new aircraft is continuously rising, aircraft are operated for longer period, and midlife update is carried out to make the aircraft competitive to contemporary aircraft. This has brought a concept of fighter generation categories created to identify major technology leaps in the historical development of jet fighters. Though there is no sound technological basis, this is more of a creation of aerospace webs and magazines. A general grouping is done based on the operational capabilities, handling qualities and pilot work load as well as the year of design. The aircraft generation is discussed below [6, 7]:

a. First-generation fighters (1945–1955)—the first-generation fighters were those built in the beginning of the jet age (World War II). These were fitted with jet but otherwise similar to earlier piston engine aircraft. They were subsonic, did not have radar and had conventional weapons like gun, dumb bombs and rockets.

b. Second-generation fighters (1955–1960)—the second-generation fighters were a class superior to the first-generation fighters as regards their speed of operation and combat effectiveness. These were fitted with radar and equipped with guided air to air missiles. This generation also took advantage of the new development of electronics in the aircraft systems.

c. Third-generation fighters (1960—1970)—the third-generation fighters were designed specifically as multipurpose fighters capable of performing both air defence and ground attack missions. McDonnell Douglas F4H Phantom, British Aerospace Harrier and MiG-23 belong to this class.

d. Fourth-generation fighters (1970–1990)—the fourth-generation (4G) fighters are high-manoeuverability multirole fighters with sophisticated avionics and weapon systems and long-range AAM. During this generation, FBW and relaxed static stability FCS concept were introduced. The advance of microcomputers in the 1980s and 1990s permitted rapid upgrades to the avionics over the lifetimes of these fighters, incorporating system upgrades such as AESA, digital avionics busses and infra-red search and track (IRST). 4+ generation fighters (1990–2000) are also sometimes used to indicate more advanced features that might be seen in fifth-generation fighters. F-16, Mirage 2000 and MiG-29 are 4G fighters, while F/A-18 Hornet, Eurofighter Typhoon and Dassault Rafale can be designated as 4+ generation fighters.

e. Fifth-generation fighters (2000 onwards)—the fifth-generation (5G) designation is used that encompasses the fighter technologies developed during the first part of the twenty-first century. The 5G jet fighters are expected to have 'pilot associates': integrated avionics and computer system capable of networking with other elements that provide the pilot with complete picture of the battlespace and situational awareness. The other features include the use of low-observable 'stealth' and high-performance airframes. Some of 5G fighters are the Lockheed Martin F-22 Raptor with USAF (2005) and the Lockheed Martin F-35 (USAF 2015) and the Chengdu J-20 with the People's Liberation Army Airforce (2017). Sukhoi SU-57 being developed for the Russian Air Force and Indian AMCA is in early stages of development.

3.3 Fighter aircraft operational profile

Combat radius or radius of action (ROA) of military aircraft refers to the maximum distance the aircraft can travel from the operating base with operational load, complete operational mission and return without refueling, allowing for reserve fuel and all other safety requirements. The thumb rule is that ROA is one third the distance an aircraft can fly on full load and total fuel. Operational mission planning is done for offensive roles to maximize the ROA without taking undue risk of enemy detection and attack. Some considerations are [1, 5]:

a. Low-level flight missions will have smaller ROA due to higher drag and fuel consumption; however, it will have low radar detection probability.

b. Aircraft with higher ordnance (weapon payload) will have low ROA.

c. A high-level mission will have higher radius of action.

d. Drop tank (D/T) increases ROA due to extra fuel, and D/T once empty can be dropped and permit aircraft to run away from the area of action.

Many offensive attack missions are normally planned a mix of hi-lo-hi mission. A typical profile is shown in **Figure 7**. The F-16 Fighting Falcon has the combat radius of 550 km (340 mi) on a hi-lo-hi mission with six 450 kg (1000 lbs) bombs.

3.4 Noncombat aircraft (NCA)

The NCA are not designed for combat as their primary function but may carry weapons for self-defence. NCA mainly operate in support roles and may be developed by either military forces or civilian derivative aircraft (CDA). NCA military applications include many different types as discussed below [1, 5]:

1. *Military transport (MT) aircraft*—both fixed-wing and rotary-wing aircraft are used for MT operation for transferring military personnel and cargo including military weapons and equipment for routine as well as urgent military operations. The operations may include rescue, tactical and strategic airlifts onto places with unprepared or semi-prepared runways. The transfers are effected to forward bases as well as areas where commercial air operations are not available.

2. *Airborne early warning and control (AEW&C)*— is an airborne platform equipped with radar, electronic intelligence (ELINT), gathered through sensors to monitor communication and radar emissions and electronic support measure (ESM) system designed to detect aircraft, ships and vehicles at long ranges and perform command and control of battle management (C2BM) in an air engagement by directing fighter and attack aircraft strikes. As the aircraft flies at altitude, the radar and other sensors have better visibility, and the onboard processor can detect, identify and track both ground and slow-moving targets. Thus, AEW&C units carry out surveillance and using the command data link act as C2BM. The AEW&C aircraft are equipped with EW system and self-protective jammers so as not to be vulnerable to counter-attack. Airborne warning and control system (AWACS) is the name of the specific system installed in the E-3 and Japanese Boeing E-767 aircraft but is often used as a general synonym for AEW&C. The Indian Defence Research and Development Organization (DRDO) is developing AEW&C on Embraer ERJ145 platform.

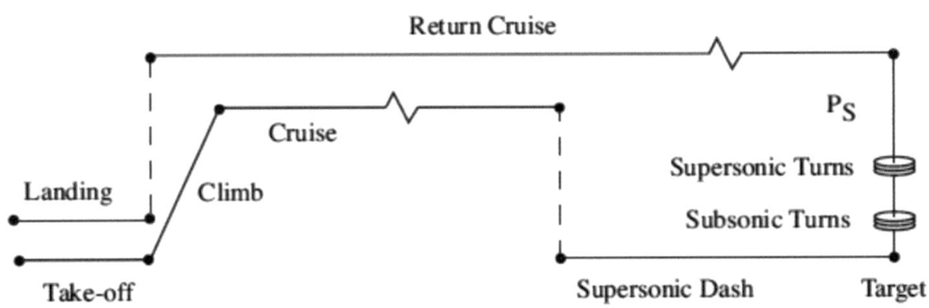

Figure 7.
Military aircraft combat mission profiles (hi-lo-hi) [5].

3. *Air-to-air fuel tanker*—air-to-air refuel procedure has become very popular in modern aviation and especially in military. In this, aviation fuel is transferred from one aircraft (tanker) to another aircraft (receiver). The aerial refueling permits the receiver aircraft to take off with greater payload (weapon, cargo or personnel) and less fuel to keep the AUW limitations. The aircraft after being airborne can be topped up with extra fuel by the tanker to increase the operational range or endurance for loiter on target location. However, both the receiver aircraft and the tanker have to be specially equipped with the precision probe and drogue system, where the receiving aircraft has fuel probe in the forward zone and the tanker is equipped with drogue to transfer the fuel. India uses IL-76 aircraft as an aerial fuel tanker.

4. *Other NCA roles*—military also uses many other aircrafts for noncombat operation. These include experimental aircraft, trainer (basic trainer and advanced trainer), reconnaissance and surveillance, search and rescue aircraft, etc.

4. Rotary-wing aircraft (RWA)

A RWA (also known as helicopter) does not have wings, but its rotor blades perform an identical role of the wings of an aeroplane. When rotor blades spin, flow of air takes place over the wings or rotor blades—hence the name rotary-wing aircraft. The magnitude of the lift can be changed by altering the angle of attack (AOA) of air over the rotor blades. The AOA is changed by mechanically increasing or decreasing the rotor blade pitch angle. For this, the pilot uses a control called 'collective' which is on his left side in the cockpit. The pilot can change the altitude of flying as well as hover the helicopter at a place by operating the collective.

Another control known as 'cyclic' is used to move the helicopter in different directions. The cyclic is operated by the pilot using his right hand. When cyclic is operated, the pitch angle of the rotor blades is changed, but it alters each blade individually by a different amount. When cyclic is operated, the total vertical force produced by rotor blades is inclined, the vertical component keeps the helicopter afloat and horizontal component moves the helicopter in the desired direction. The helicopter cyclic control system is designed in such a way that when the cyclic is moved forward, the helicopter moves forwards, when it is moved sideways the helicopter moves sideways and when cyclic is moved aft the helicopter moves backwards (which a fixed wing aircraft cannot do). The third helicopter control is the yaw pedals. These alter the pitch angle of the tail rotor—the small rotor at the end of the helicopter. Doing this enables the pilot to turn the helicopter either left or right (yaw) [8].

4.1 Military application of helicopters

The helicopter is a very useful machine for military applications. Due to its design features, helicopters can hover at a point, fly at very low altitude and above all take off from the land at any place it desires. Military helicopters are also installed with protective armor/windshield against bullets. Some of the typical applications are:

1. *Attack helicopter*—attack helicopters are used in the antitank and close support roles. These helicopters are equipped with antitank-guided missiles, guns and rockets. To enable them to find and identify their targets, some modern attack helicopters are equipped with very capable sensors such as a millimetre wave radar system.

2. *Antisubmarine warfare (ASW) helicopters*—helicopters used for ASW roles can be ship based or land based and are equipped with winch for use of dunking sonars to detect, identify, locate and track the target submarines. To destroy submarines, both torpedoes and naval mines are used, launched from air, surface and underwater platforms. It is also used to protect friendly ships.

3. *Observation helicopter*—the ability to both manoeuvre and hover at one location makes the helicopter ideal for reconnaissance. Modern observation helicopters are equipped with optical sensor systems. These include low-light video and forward-looking infra-red cameras. Often, these are mounted on a stabilized platform along with multifunction lasers capable of acting as laser range finder and targeting designators for weapon systems. Advanced observation helicopters are equipped with sensor suits capable of providing terminal guidance to ATGWs, laser-guided bombs and other missiles and munitions fired by other aircraft.

4. *Transport helicopter*—transport helicopters are used for transporting personnel (troops) and cargo in support of military operations. The benefit of using helicopters for these operations is that personnel and cargo can be moved to and from locations without requiring a runway for takeoffs and landings. Cargo is carried either internally or externally by slung load suspended from cargo hook. In difficult terrains where helicopters cannot land, personnel may be picked up and dropped off using rescue hoists or special rope lines, while the aircraft hovers overhead. Helicopters are also used in air assault, where customized assault force assembled on the pickup zone with their equipment is transported to a landing zone (LZ). The use of helicopter in air assault permits transport and lands a large number of troops and equipment in a relatively short time, near the LZ. The transfer is done in stages and sequentially thus supports the assault by continually resupplying the force during the operation.

5. *Medical evacuation (MEDEVAC) helicopter*—MEDEVAC is the timely and efficient movement and en route care provided by medical personnel to wounded personnel being evacuated from a battlefield or to injured patients being evacuated from the scene of an accident to receive medical facilities. Helicopters are used as air ambulance.

6. *Combat search and rescue (CSAR) operation helicopters*—search and rescue (S&R) operation includes search and supply provision to people who are in distress or imminent danger. Normally S&R is carried out by specially trained volunteers. CSAR are search and rescue operations that are carried out within or near the combat zone during war. A CSAR mission may be carried out by a task force of helicopters, ground attack aircraft, aerial refueling tankers and an airborne command post from which a commander controls and organizes his forces.

7. *Utility helicopters*—a utility helicopter is a multipurpose helicopter. A utility military helicopter can fill roles such as ground attack, air assault, military logistic, medical evacuation and troop transfer. Hindustan Aeronautics Limited (HAL), India, is developing a light utility helicopter for the services of the Indian Armed Forces.

5. Unmanned aerial vehicle (UAV)

An airborne vehicle capable of being flown without a pilot on board is termed as pilotless or unmanned aerial vehicle (UAV). UAVs are broadly in two types, namely, remotely piloted vehicle (RPV) and autonomous and preprogramed air vehicle [9, 10].

5.1 Application of UAV

UAV has the potential to serve varieties of application both in civil and military uses. These include border and port security, homeland surveillance, scientific data collection, cross-country transport and telecommunication services. Many non-aviation business applications include:

1. Aerial photography, film, video, still, etc.

2. Agriculture crop monitoring and spraying

3. Conservation, pollution and land monitoring

4. Electricity companies for their power line inspection.

5.2 Military UAV

UAVs used for armed forces, police, border security and coast guard are considered owned by the state commonly known as military UAV. These UAVs are under direct control of the government of the state, and the state is directly responsible for safety of operation and third-party damage in the event of UAV failure. This direct control of operations is a significant advantage of operation in segregated space at the same time accepting a safety of operation by imposing operational restriction to compensate for uncertainties over airworthiness.

5.3 Military application of UAV

UAV in military is used as a force multiplier for carrying out very dull, dirty missions that are long and considered very hazardous and risky for manned flight. Some of these tasks are round-the-clock reconnaissance, surveillance and target acquisition (RSTA); nuclear, biological and chemical (NBC) war; and combat search and rescue (CSAR) operation. In offensive role UAV can be employed for arm dropping as well as suppression of enemy air defence (SEAD).

Unmanned combat air vehicle (UCAV) is used for offensive war application. These are designed to possess:

a. High accuracy and probability to 'strike the target capability'

b. All weather operation

c. High speed and manoeuverability for war zone penetration

d. Autonomous programme with rerouting facility

5.4 UAV vs. manned aircraft

UAVs are aircraft within the meaning of the International Civil Aviation Organization (ICAO) Convention, 1944 (§8 of Chicago Convention). As per §3 of Chicago Convention, the ICAO rules do not apply to state (military) aircraft. The airworthiness and safety associated with flying UAV in segregated area (military airspace) are therefore state responsibility. However, special political agreement will be required if the military UAVs fly over the territory of another state (§8 of the ICAO). Further, §20 of the ICAO requires each UAV to bear registration and nationality mark, and §8 requires special authorization of state (military) for UAV to fly over its area. And above all article, §33 requires UAVs to have current certificate of airworthiness.

5.5 Integration of UAV into national airspace

Though UAVs are employed in segregated areas at present, there is a consensus view within the aerospace industry that the time is ripe when both manned and unmanned aerial vehicles will share the common airspace. Thus, the process of integrating the manned aircraft and UAV in the national airspace has to be accepted and regulated. This needs to regulate the operation of UAV on the one hand, and at the same time, the UAVs themselves have to be certified to be airworthy by regulatory organizations. Further, regulatory activities of air traffic management for integrating UAV in non-segregated airspace operations have to be considered:

a. *Certification consideration for UAV*—as technology advances, legislations need to be enacted to avoid proliferation. Existing regulatory procedure for manned flight cannot readily be made applicable to UAV. Therefore, suitable developments or amendments in UAV design process need to be initiated to make them acceptable in the common airspace. Besides the specialized infrastructure facilities in the air traffic management also have to be developed to fly UAV and manned aircraft in the same airspace.

b. *Type certification standard*—UAV-type certification standard should in addition to manned aircraft requirement consider the following:

- Emergency recovery capability
- Communication link and link loss criticality
- Level of autonomy
- Human-machine interface (UAV pilot is deprived of the physical senses and feeling of flying as in manned aircraft pilots)
- Ground control system and launch and recovery system should be subjected to functional hazard analysis and accordingly certified.

5.6 Collateral damage

UAV operation has a great risk of collateral damage in the event of UAV failure. The regulatory principles thus aim to take all precautions to reduce the risk of collateral damage. The considerations are:

a. Design for low collateral damage—by designing with high precession and accuracy in striking the target using high-accuracy sensor and doing proper mission planning to hit the target and not the structure/people around (third party).

b. Approach for damage prevention—the damage is proportional to the kinetic energy (KE) during the impact. The KE depends on the mass and velocity at the time of impact. The V_{impact} will be based on the cause of the failure, namely, engine or control failure. The general rule is to assume V_{impact} = 1.3 times the V_{stall} for the case of unpremeditated descent or engine failure and V_{impact} = 1.4 V_{stall} in the event of loss of control.

c. Design safety consideration—the basic safety criterion should be that catastrophic failure conditions must be extremely improbable and, for all other failure conditions, the probability of occurrence of the event should be inversely proportional to the damage potential of the failure. From airworthiness point of view, the risk to third parties on the ground would become the most severe risk to be minimized. Depending on the 'hit' probability on the ground (a function of population density and UAV lethal area), some operational limitation with regard to 'overflown zone' may be imposed.

5.7 Emergency recovery system

Flight termination systems (FTS) should be installed as a means of recovering UAVs from system failures. FTS can be an automatic flight guidance system which navigates the aircraft to a suitable location and completes a normal landing or devices which bring the aircraft down immediately, e.g. by deployment of a parachute, e.g. JAR 1309 (ballistic recovery system (BRS)). It is noteworthy that BRS has been fitted for some time to certain manned civil aircraft, notably microlights. The current CAA policy on such systems is that they may be installed on a 'no hazard, no benefit' basis only. A parachute may be fitted if desired, but it is not to be relied upon to prevent an accident. Applicants for the approval of aircraft embodying FTS have to show that the system is protected from inadvertent operation or that the consequences of inadvertent operation are acceptable.

6. Airworthiness and flight safety

6.1 Airworthiness

Airworthiness in a simple term can be defined as 'fit to fly'; however, in an actual sense, it is defined as 'demonstrated capability of an airborne store to perform satisfactorily and fulfil the mission requirements, throughout the specified life in the prevailing environments with acceptable level of safety and reliability'. The 'Federal Aviation Administration' (FAA) declares an aircraft is airworthy if it conforms to its type design and if it is in a condition for safe flight. The first part of the FAA definition describes the airworthiness requirement for 'certification', and the second part of the definition refers to the 'continued airworthiness' which regulates the repair, maintenance and operation throughout the life span of the aircraft [11–27].

6.2 Airworthiness and flight safety in military aviation

The phrase 'acceptable level of safety' in the definition of airworthiness is a complex consideration as absolute safety is hypothetical and can be achieved only at infinite cost. Therefore, the airworthiness standards have to balance between safety concerns on the one hand and the cost and practicability from design and manufacture point of view on the other.

In aviation, safety may be defined as freedom from death, injury or damage to people on board and loss property and life on ground (accident). Safety of any flying effort or machine would depend primarily upon, whether we are operating below or above a particular level known as 'risk threshold'. The risk threshold is the level of risk beyond which accidents are inevitable. It also must be appreciated that this 'risk threshold' is not a stationary one and it keeps varying based on the role, function and a host of other associated factors. It needs therefore to be reassessed under each changing scenario. Airworthiness control is to minimize the risk and maximize the effectiveness. All the airworthiness standards, military or civil, whether that of the USA, Europe or Russia, have a common point of reference which is that an inverse relation should exist between probability of occurrence of an event and the degree of hazard inherent in its effect.

For military aircraft in the US Department of Defense (DoD) document, Mil-STD-882 defines the safety requirements during design of an airborne stores. The decision matrix of system safety as per Mil-STD-882 is shown in **Figure 8**. The safety requirement of any failure event is based on the hazard index of the failure which is defined as the product of the probability of the failure and the damage consequent of the failure event. The damage consequences can range from 'catastrophic' (loss life, aircraft and property) to negligible (minor inconvenience). The frequencies of occurrences are grouped under 'frequent' (1 in 10 h of flight) to 'extremely improbable' (1 in 10^7 flight hours).

6.3 Flight safety directorate

The operational branches of both military and civil aviation are expected to have effective 'flight safety directorates'. The primary responsibility of the flight safety directorate is to estimate the risk threshold under all dynamic condition, take appropriate measures and ensure that operational risk does not exceed this value. The basic aim of the flight safety studies is to ensure that the chances of achieving the mission tasks be optimal, while operational risks are minimal. In military aviation, the flight safety directorate has a very complex duty to perform. On the one hand, the military training must give a high level of exposures to possible war scenario and threats,

Criticality ↑						
	Catastrophic I	HAWR	HUD	HUA	HUA	HUA
	Critical II	HAWR	HUD	HUD	HUA	HUA
	Marginal III	HAWR	HAWR	HUD	HUD	HUA
	Negligible IV	HA	HA	HA	HAWR	HAWR
Frequency →		Level E Improbable	Level D Remote	Level C Occasional	Level B Probable	Level A Frequent

Hazard Un Acceptable (HUA)
Hazard Un Desirable (HUD)
Hazard Acceptable With Review (HAWR)
Hazard Acceptable (HA)

Figure 8.
Mil-STD-882: design safety decision matrix [14].

while it must ensure that the high level of risks are to be avoided in peace time within the stated training syllabi. This is because the accidents have very deleterious effects on the morale of the flier. During the war time, however, task achievement is paramount, and hence risks are to be taken even at high degrees if the operational requirements dictate. The main purposes of the flight safety studies are therefore to:

a. Identify and minimize those risks which may contribute accidents

b. Avoid very high cost of losses and damages

c. Identify all risk hazards real or potential at all levels and all phases of flying

d. Risk thresholds are dynamic, and they need to be reassessed under all conditions and make necessary readjustments

7. Military aircraft design and certification

7.1 Military aircraft design consideration

All modern military aircraft are designed to perform multiple missions. This is inherent in the design of such aircraft, for example, the A400M military transport aircraft is designated as a tactical airlifter with strategic capabilities and can also be configured to perform long-range cargo and troop transport, medical evacuation, aerial refueling and electronic surveillance missions. *Military aircraft* is capable of operating from semi-prepared runways under varied climatic and environmental conditions [1, 13, 28].

Military aircraft design requirement is driven by operational capabilities derived from threat perception and defence preparedness strategy of a country, whereas commercial aircraft design and development are driven by the market forces. A comparative study of the design *considerations for civil and military* aircraft is shown in **Table 1**.

7.2 Certification of aircraft

Certification is a process of evaluation and documentation of compliance of a product to its specified requirements and declares it 'safe to fly'. It is a third-party assurance to the user that the product has been designed, developed, evaluated and produced in such a manner that its quality, reliability and integrity meet the requirements. An aircraft can really be considered airworthy when [13]:

a. Its type has been designed and certified meeting design standards.

b. It has been manufactured by an approved organization as per type design.

c. It has been maintained by qualified people as per approved system and inspected in accordance with all applicable airworthiness directives.

d. Further, on compliance of (c), no significant defects have been found and not rectified.

7.3 Military certification

There is a difference in concept of civil and military aircraft certification. While military certification has a general concern of airworthiness as well as

Objectives	Civil aircraft design	Military aircraft design
Requirement	Independent market survey	Military staff requirement
Certification goal	Civil-type certification used by airlines	Providing national security Issue operational clearance
Accepted failure tolerance/flight	Level 2: $p < 10^{-4}$ Level 3: $p < 10^{-6}$	Level 2: $p < 10^{-2}$ Level 3: $p < 10^{-4}$
Design safety	As per SAE ARP 4671; permitted failure rate 1 in 10^9 flying hours	As per Mil-STD-882; permitted failure rate 1 in 10^7 flying hours
Avionics architecture	Better FCS, navigation, reduce pilot workload, increase payload	High performance, navigation and weapon aiming, secure COM and reduction in pilot workload
Use of advanced technology	Only certified material/technology	Cutting edge technology; certified/ under development
Stealth tech/EW protection	Not considered	Most essential consideration
Design life	High economic life	High maneuverability, extreme operating envelope

Table 1.
Design consideration—Civil and military aircraft [13].

rules for design and performance evaluations, each country has its own rule to 'self-certify' its state aircraft as airworthy and compliant to some specified and controlled airspace performance requirements. While certifying a military aircraft, the operational risk and operational process are defined for each type of aircraft. Even though the purpose of airworthiness control is the same, the civil and military certification differs on the fact that [13]:

a. Governments can 'self-certify' their state aircraft.

b. Operational risk is defined and accepted by the service.

c. Military certifications differ on the degree and coverage of the evidences needed. This is in general limited by contract, budget, lack of past legal liability and aircraft type and legacy.

d. Acceptance of specific tasks and the risk levels can vary with aircraft purpose and type.

Military certification also differs from civil procedure due to the fact that military certification looks for induction of the aircraft into service use. The certification tasks also include vehicle performance evaluation and system qualification for induction into service. The induction clearance is given through issue of 'release to service' document. The general certification procedure follows the following routes:

a. Approval builds up in a building block method.

b. The first-level clearance includes qualification of component and equipment through lab-level test.

c. The second-level clearance includes demonstration of systems' performance through aircraft on ground tests and system simulations.

d. Finally the aircraft-level performance is evaluated through flight test programme.

e. A compliance document is prepared capturing the evidences generated through analysis, lab test, aircraft inspection, system integration test and flight tests.

f. Finally a 'release to service document' is issued.

7.4 Military certifications in different countries

Military aviation is an important factor in security and defence preparedness of each state. All military airworthiness activities are conducted and regulated on a national basis, and in general most military authorities have not published military airworthiness design standards for an acceptable level of safety. In general for the military airplanes, the military specification and documents nominate some specified elements like:

a. Handling qualities

b. Weapons, ammunition stores and self-defence suites

c. Specific operations in wartime

d. Military role and mission and tasks

While the principle remains the same, the practices evolved in different countries over a period of time, differ from each other to a varying degree. The range of control over all the activities in design, development evaluation, and testing varies from total control to delegated system of working, with emphasis on contracts and penalties for shortfalls in performance, time and cost overruns [13, 14, 18–27, 29].

7.4.1 Military certification procedure in the United Kingdom

With effect from 1 April 2010, the Secretary of State (SofS) for Defence created the Military Aviation Authority (MAA) by charter as the single independent regulatory body for all defence aviation activities in the United Kingdom. The Director General of the MAA (DGMAA) is accountable to SofS, for:

a. Providing airworthiness regulatory framework

b. According airworthy clearance through certification

c. Approving aircraft inspection process for the acquisition

d. According airworthiness assurance of all air systems held in the inventory of defence aviation environment

Accordingly DGMAA prepares and releases regulatory publications (MRP) and has the authority to issue them on behalf of the SofS. The MAA has published the document 'RA 1500 – Certification of UK Military Air Systems'. This document has now been superseded by RA 5800 and RA 5820 [19–21].

7.4.2 Military certification procedure in America

According to the US Air Force (USAF) Policy Directive No. AFPD 62-6 on 11 Jun 2010 issued by the Secretary of Defense, the USAF is responsible for assuring the airworthiness of all the aircraft which it operates. The directive establishes policies for formal airworthiness evaluations to ensure that AF-operated aircraft are airworthy over their entire life cycle and maintain high levels of safety. This policy is implemented through USAF Instruction AFI 62-601, dated 15 Jun 2010 and supplemented on 12 May 2011. According to this AFI, a Technical Airworthiness Authority (TAA) and an AF Airworthiness Board (AFAB) have been created in the AF Materiel Command (AFMC) to provide independence in airworthiness evaluations. AFAB is chaired by the TAA. AFAB defines the requirement for *design-based (DB) airworthiness certification* and also provides *non-design-based (NDB) special flight release* of aircraft when design-based certification is not possible. AFAB provides acceptance of FAA certifications, evaluations and inspections and disestablishes the Airworthiness Certification Criteria Control Board (AC^3B). Based on the technical evaluations, the Project Manager (PM) proposes one of the two possible alternatives of DB certification or NDB special flight release. The former one is the preferred approach, while the latter is chosen 'by exception' on unique aircraft or situations.

7.4.2.1 Design-based airworthiness assessment

The DB certification is carried out in stages of design evaluation, issuance of military experimental flight release and finally issuance of military-type certificates (MTC). Issuance of MTC indicates that aircraft design documentation accurately defines the configuration which meets the certification basis and the aircraft design is in compliance with requirement.

7.4.2.2 Non-design-based airworthiness assessment

A Non Design Based (NDB) assessment is conducted when it is found by the TAA that a DB airworthiness certification cannot reasonably be accomplished, but there is a compelling military need to operate the air system. On successful conclusion of NDB evaluation, the TAA may issue a special flight release. The NDB special flight releases process identifies and assess the inherent risks of operating these aircraft and the services formally acknowledge these risks during their flight operations.

7.4.2.3 Commercial derivative aircraft (CDA)

The USAF prefers FAA-type certification for newly developed military transport and CDA for USAF operation, when it is found that the criticality of military usage is no severe than the FAA-certified flight envelope and operational environment. FAA Form 8130-2 or 8130-31 can be used for FAA-type certification. CDA require the issuance of a MTC by the TAA. For this, FAA TC is used for the basic aircraft, and a compliance analysis is carried out with the approved military certification basis for items not covered by FAA (e.g. EW suits or other military appliances). MIL-HDBK-516 is to be used to define applicable military airworthiness certification criteria. CDA used in the USAF are to be maintained as per AFI 21-107: maintaining Commercial Derivative Aircraft.

7.4.3 Military airworthiness of Australia

In 2014, the Chief of Defence, Australian Defence Forces (ADF) and Secretary of Defence decided that the procurement and maintenance of all aviation fleet of

DOI: http://dx.doi.org/10.5772/intechopen.87087

ADF be regulated to meet the requirement of Australian Defence Safety Objectives. DoD, Australia, issued Australian Air Publication AAP 7001.048, 'Defence Aviation Safety Program (DASP) Manual' on 30 Jun 2014. As per the above publication, the Chief of Air Force will be the 'Defence Aviation Authority'. He will be responsible for creating and implementing DASP. He will be supported by Deputy Chief of Air Force (DCAF) as Operational Airworthiness Regulator (OAR) and Directorate General Technical Airworthiness (DGTA).

7.4.4 Military airworthiness in Russia

The military aviation design and development and certification were carried out by various aviation design bureaus and manufacturing complexes. Military forces decide the operational requirements and release design specification. The aviation design bureaus and the manufacturing units carry out necessary designs to meet the requirements. The design bureaus take support from the state research institutes on aerodynamics and other aircraft systems. In fact the design bureau creates a number of alternate designs. The designs are evaluated, and the chosen design is then assigned to one or more manufacturing complexes. During D&D and manufacture of the aircraft, state standards GOST are to be used. The production and quality system 'Oboronsertifika' followed by the defence industries are similar to international quality standards like ISO, AS9100 [13].

7.4.5 Military airworthiness and certification in India

The Indian military airworthiness certification process has been modeled after the earlier British system. The system is based on concurrent design and clearance leading to eventual certification. This approach was adopted since the certification authority, viz. the Chief Resident Engineer (CRE), and the inspection authority Chief Resident Inspector (CRI) were co-located with defence public sector Hindustan Aeronautics Limited who is responsible for design and development (D&D). D&D milestones are agreed between the design and certification authorities, and the design is reviewed for safety and airworthiness by a team of experts. At appropriate stages, test procedures are examined and approved, and tests are carried out. The test results are reviewed for acceptance, redesign or retest. The CRE and CRI organizations have been changed to CEMILAC and DGAQA; however, the inspection and design certification procedure have practically remained the same.

However, this process conceptually differs from those followed in the United States and Europe, where both D&D and manufacture are delegated to approved design/production organization. The Defence Project Managers (PM) monitor the project progress. At mutually accepted major milestones, reviews are carried out as per the agreed documentation of the contract. The government, through the certification provision, holds the authorized personnel within the firm responsible for the airworthiness certification of the aircraft. In contrast, in India, CEMILAC and DGAQA interact with the D&D team on a day-to-day basis and carry out spot checks to identify design/production deficiencies during the D&D stage [3, 4, 13, 29].

MOD Document DDPMAS 2002 (Procedures for Design Development of Military Aircraft and Airborne Stores) guidelines are as follows. DDPMAS 2002 volume 2 is used as a procedure for certification of airborne software. These documents also lay down airworthiness assurance procedures during manufacture, overhaul and upgrade of military aircraft. The military airworthiness functions are shown in **Figure 9**, and aircraft certification procedure is shown in the block diagram in **Figure 10**.

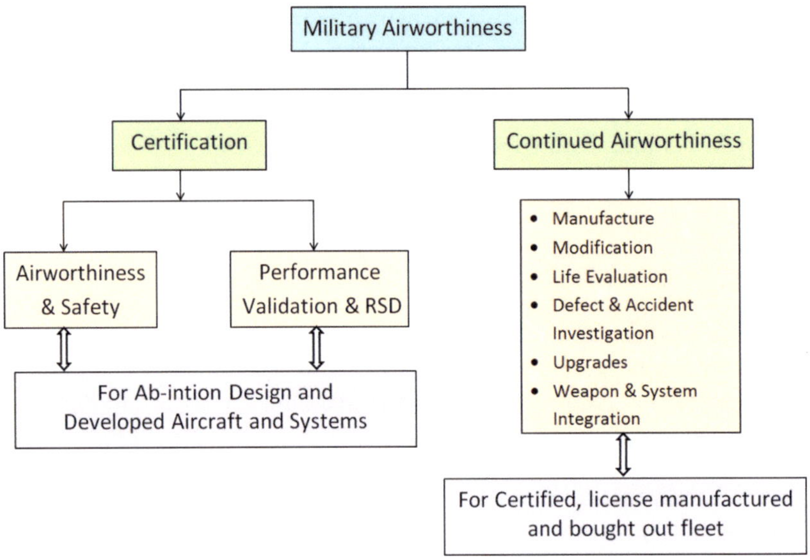

Figure 9.
Airworthiness functions [13].

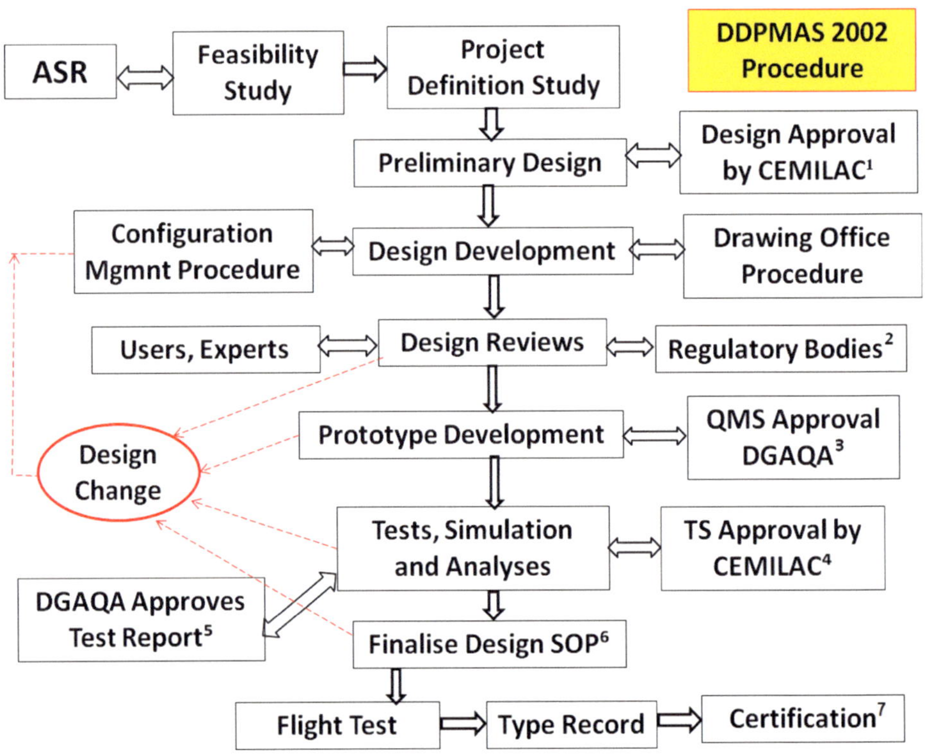

Figure 10.
Certification procedure [13].

8. Conclusion

In the modern warfare, air power plays a very significant role in both offensive attack role and defending the country's land and water space. Judicious mix of various types of combat and noncombat aircraft increases the war power of both army and navy. Competition to develop more effective air weapon is thus never-ending. Huge research effort is being spent on development of sophisticated military aircraft with precision navigation, weapon aiming and targeting along with improvement of air armament lethality.

Aviation is also associated with various safety issues. While in civil aviation, ICAO regulations take care of aviation safety issues of the member states, military aviation safety is controlled by each state according to their own military doctrine. To make military aviation safe, the authorities must exercise effective regulatory control during design and development, operation and maintenance of the military aviation assets.

Author details

Kanchan Biswas[1,2]

1 CSIR-NAL, Bangalore, India

2 CEMILAC, DRDO, Bangalore, India

*Address all correspondence to: kanchan.biswas@rediffmail.com

References

[1] Kundu AK. Aircraft Design. New York: Cambridge University Press; 2010

[2] Houghton EC, Carpenter PW. Aerodynamics for Engineering Students. 5th ed. Oxford, UK: Butterworth-Heinemann; 2003. DOI: 978-0-7506-6948-1

[3] Nagraj K, Kalyanam VK, Annamalai SP. Reference Book on Airworthiness and Certification of Military Fixed Wing Aircraft. IDST-B Report. Bangalore: CEMILAC; 2005

[4] Biswas K. Lecture notes: Aerodynamics. Adjunct Faculty. Bangalore, India: IIAEM, Jain University; 2016

[5] Hull DG. Fundamentals of Airplane Flight Mechanics. Berlin Heidelberg: Springer-Verleg; 2007. DOI: 13-978-540-46571-3

[6] Hebert AJ. Issue Brief: Fighter Generations. Arrington, VA: Air Force Magazine; 2008. p. 32

[7] Tirpak JA. The Sixth Generation Fighter. Arrington, VA: Air Force Magazine; 2009. pp. 38-42

[8] Seddan T. Basic Helicopter Aerodynamics. London: BS Professional Book; 1990

[9] CAP-722: Unmanned Aircraft System Operations in UK. 6th ed; Civil Aviation Authority of UK. 2015

[10] ICAO Circular AN/190. Unmanned Aircraft System (UAS). 2011. DOI: 978-92-9231-751-5

[11] Gratton G. Initial Airworthiness, Determining the Acceptability of New Airborne Systems. Switzerland: Springer International Publishing; 2015

[12] De Florio F. Airworthiness: An Introduction to Aircraft Certification. UK: Elsevier Publication; 2006

[13] Biswas K. Military airworthiness and certification: Global scenario. In: Proceeding of the International Conference on Modern Research in Aerospace Engineering, MARE-2016. Singapore: Springer; 2018. pp. 315-331. DOI: 10.1007/978-981-10-5849-3-31. Available from: https://link.sprnger.com/chapter/10.1007

[14] Mil-Std-882D: Standard Practice for System Safety Assessment; USA: DoD. 2000

[15] SAE. ARP 4761: Guidelines and Methods for Conducting Safety Assessment Process on Civil Airborne Systems and Equipment; SASE; 1996-2012

[16] Raheja DG. Assurance Technologies: Principles and Practices. McGraw-Hill Engineering and Technology Management Series, Inc; 2006

[17] Airworthiness Manual. 3rd ed. International Civil Aviation Organization. Doc 9760; 2013

[18] MAA 01: Military Authority Regulatory Policy; Military Aviation Authority, Govt. of UK; Issue-2. 2012

[19] RA 1500: Certification of UK Military Registered Air Systems. Regulatory Article, Last Issue. Govt. of UK: Military Aviation Authority; 2016

[20] Regulatory Article 5800: General requirements – project teams and organisation (MRP 21). Govt. of UK: Military Aviation Authority; 2018

[21] Regulatory Article 5820: Changes in Type design (MPR 21, Sub Part D). Govt. of UK: Military Aviation Authority; 2018

[22] Air Force Policy Directive 62-6: USAF Airworthiness; DoD, USAF. 2010

[23] Air Force Instruction 62-601: USAF Airworthiness; DoD, USAF. 2011

[24] Air Force Policy Directive 62-4: Standards for Airworthiness for Passenger Carrying Commercial Derivative Transport Aircraft; DoD, USAF. 1988

[25] FAA Order No. 8100.101: Type Certification Process for Military Commercial Derivative Aircraft; US Department of Transportation National Policy. 2007

[26] Mil Std-516B: Airworthiness Certification Criteria; US DoD. 2005

[27] Australian Air Publication 7001.048. Defense Aviation Safety Program Manual; Department of Defense, Government of Australia. 2012

[28] Roskam J. Airplane Design. Vol. I–VII. Ottawa, Kansas: Roskam Aviation and Engineering Corporation; 1985

[29] DDPMAS. Design, Development and Production of Military Aircraft and Airborne Stores. Department of Defense Production and Supplies, Ministry of Defense, Government of India; 2002

Chapter 2

Scientific Knowledge of Spanish Military Engineers in the Seventeenth Century

Josep Lluis i Ginovart

Abstract

The catenary arches were used in Spanish Art Nouveau architecture by Antoni Gaudí (1852–1926). The theory of the chain, in the shape of a hanging collar, was proposed by Robert Hooke (1676) and used by Christopher Wren in Saint Paul's dome (1675). British school modern mechanic theory was introduced in Spain by Spanish Bourbonic military engineers and also by the Catholic Scottish and Irish families during the eighteenth century. The assessment of some drawings of gunpowder warehouses, found in the collection of *Mapas planos y Dibujos* (MPD) of the General Archive of Simancas (*Archivo General de Simancas*, AGS) (AGS 2014), has revealed the use of the chain theory in Miguel Marín's projects for Barcelona (1731) and Tortosa (1733) and Juan de la Feriére ones in A Coruña (1736). A built evidence has also been found: the Carlón wine cellars in Benicarló, built by the O'Connors family from Ireland (1757). The analysis of these examples proved the theory of the chain arrival to Spain during the first half of the eighteenth century.

Keywords: catenary, gunpouder warehouse, military engineering, geometry, Robert Hooke

1. Introduction: the catenary arch in Spain in the nineteenth and twentieth century

Catenary arches are one of the main features of Art Nouveau architecture in Spain. Their shape is based on the modern theory of masonry arches. This theory was developed during the nineteenth century and claimed the work of Antoni Gaudí (1852–1926) as its main exponent [1]. Architect Cèsar Martinell i Brunet (1888–1973) built the so-called wine cathedrals (1918–1924) which were built by the Commonwealth of Catalonia (1907–1925).

These buildings in the nineteenth-century style named as Noucentisme may be regarded as the last ig cluster of constructions featuring Catalan masonry [2]. The catenary arches designed and built by Cèsar Martinell belong to the Catalan modernist architecture Antoni Gaudí i Cornet (1852–1926). Thus, in a hanging chain, any inward pulling force is matched by an equal outward pushing force. Martinell knew Gaudi's work and inherited his techniques. This is evidenced by his many writings on Gaudí [3], whom he met during a visit to the Sagrada Familia church in 1915, when Martinell was about to complete his studies at the Barcelona School of

Architecture. From then on, Martinell joined an exclusive group of disciples who learned a way of doing architecture aside from university teachings [4].

For Martinell, Gaudí was a much more interesting lesson of life and architecture than most of the teachings given at university. Gaudí's words became architectural when the very statement of the scientific truth and the procedures he himself invented were able to explain problems and geometric concepts which remained unclear in the classrooms of the school of architecture. Antoni Gaudí's theory of structures relies on the strength of geometry, in particular on the strength of the parabolic and catenary shapes. The construction technique used by Cèsar Martinell stems from the methods applied by Gaudí to calculate the geometric shapes of vaults and arches [5] (**Figure 1**). Other architects have written about this view; see, for instance, the lecture entitled *La fábrica de ladrillo en la construcción catalana* (1900), by Josep Domènech i Estapà (1858–1917). In this lecture it is claimed that the parabolic and catenary shapes are the lines of equilibrium in a system of evenly distributed loads, where the parabolic shape relates to the horizontal projection and the catenary shape relates to the arch length [6].

Otherwise, these concepts were introduced in the formation of architects through the Escuela Especial de Arquitectura de Madrid (1844). The work *Traite Theorique et Pratique de L'art de bâtir* (1802–1817) by Jean-Baptiste Rondelet (1742–1829) exposes the methodology to lay out catenary arches by means of the theory of the chain and another complicated procedure [7]. In addition, a treatise by John Millington (1779–1868) was also used in architecture schools. It was translated as *Elementos de Arquitectura* (1848) and contained Hooke's theory and the layout of the catenary [8]. Juan Torras i Guardiola (1827–1910) developed the scientific basis for the calculation of these structures in the Barcelona School of Architecture (1875) [9].

1.1 The curve of equilibrium

The theory of the chain, in the shape of a hanging collar, was proposed by Robert Hooke (1635–1703) at the end of his treatise *A description of Helioscopes, and Some Other Instruments* (1676). Hooke presented a solution that would be revealed as "Ut pendet continuun flexile, sic stabit contiguum rigidum inversum" [10]. The

Figure 1.
Cesar Martinell sketch of transversal section El Pinell de Brai [COAC H101I-6-Reg 2502].

awareness about the shape of the catenary was applied by Christopher Wren (1632–1723) in the dome of San Pablo (1675), with the collaboration of Robert Hooke in the design [11].

Simon Stevin (1548–1620), in *De Beghinselen der Weeghconst* (1586), had previously proven the law of equilibrium of a body on an inclined plane. There we can see a hanging cable which has the shape of a catenary [12]. Despite the evidence provided by the figure, there was not any mathematical approach to catenaries. That is why Jakob Bernoulli (1654–1705), in *Acta Eruditorum* (1690), issued a challenge to the mathematical community to solve this problem [13]. The solution was published in *Acta Eruditorum* (1691) by Johann Bernoulli (1667–1748) with the title "Solutio problematis funicaularii" [14] and also by Chistiaan Huygens (1629–1695) with the title "Dynastae Zulichemii, solutio problematis funicularii" [15].

The mathematic equation of the catenary would be formulated some years later by David Gregory (1659–1708) and published in the *Philosophical Transactions of the Royal Society* (1697). Gregory affirms that the catenary is the real shape of an arch, because if these can sustain themselves, it is because a catenary can be drawn in its section [16]. James Stirling (1692–1770), in the *Lineae Tertii Ordinis Neutonianae* (1717), compiled the ideas of the English school building a catenary with hanging spheres, to simulate the behavior of a constructive element [17]. This solution inspired the analysis by Giovanni Poleni (1683–1761) in the *Memorie Istoriche della Gran Cupola del Tempio Vaticano* (1748) [18], who developed a methodology similar to Stirling's, to understand the breaking of the vault of the San Pietro Basilica [19].

In Spain, the development and application of this theory take place in the context of the Mathematics Academy of Barcelona (1720). The work of Bernard Forest de Bélidor (1698–1761) is the main reference of the curve of equilibrium theory. In *La science des ingénieurs dans la conduite des travaux de fortification et architecture civile* (1729), Book II, Chap. III, Prop. V, Bélidor sets out the curve that must be given to a vault, so all its parts weigh the same and stand in equilibrium [20], and as a result its curve will have the shape of a catenary. And so he determines, for military constructions, up to five types of different vaults: rounded, tiers-point pointed, elliptical drawn as a segmental arch, the flat ones, and the derived forms of the catenary [21]. In addition, the work *De la poussée des voûtes*, (1729), by Pierre Couplet (+1743), mentions the *chaînette*, the hanging chain, as the best of all shapes for the construction of vaults. He also says that to build this vault, every part of the hanging rope has to be loaded with the proportional weight of the construction, so the resultant curve will be the one to be used [22].

2. The libraries of the military engineers of the eighteenth century

On 13 January 1710, King Philip V appointed Jorge Prosper Verboom (1665–1744) as engineer in chief. Verboom was a disciple of Sebastián Fernández de Medrano. Together with Alejandro de Retz (c. 1660–c. 1732), chief engineer of the Catalan region, Verboom was the link with the former academy in Brussels. Mateo Calabro was appointed head of the Mathematics Academy of Barcelona (1720–1738), and he had profound disagreements with Jorge Prosper Verboom with regard to the training program. Therefore, in 1738 Verboom offered the head position of the academy to Pedro de Lucuze y Ponce (1692–1779), who held that position until his death. Among other duties, the academy had to build a collection of scientific works which would be used as reference texts for military training. This bibliographic interest led Vicente García de la Huerta (1734–1787) to publish *Bibliotheca Militar Española* (1760), a collection of the most important military engineering treatises which had been written between the sixteenth and the eighteenth centuries [23].

Some of the most relevant texts available for military engineers in the libraries are *L'architecture des voûtes* (1643) by François Derand (1588–1644) (**Figure 2**), in the library of Jorge Prosper Verboom (1665–1744) [24], and the *Treatise on*

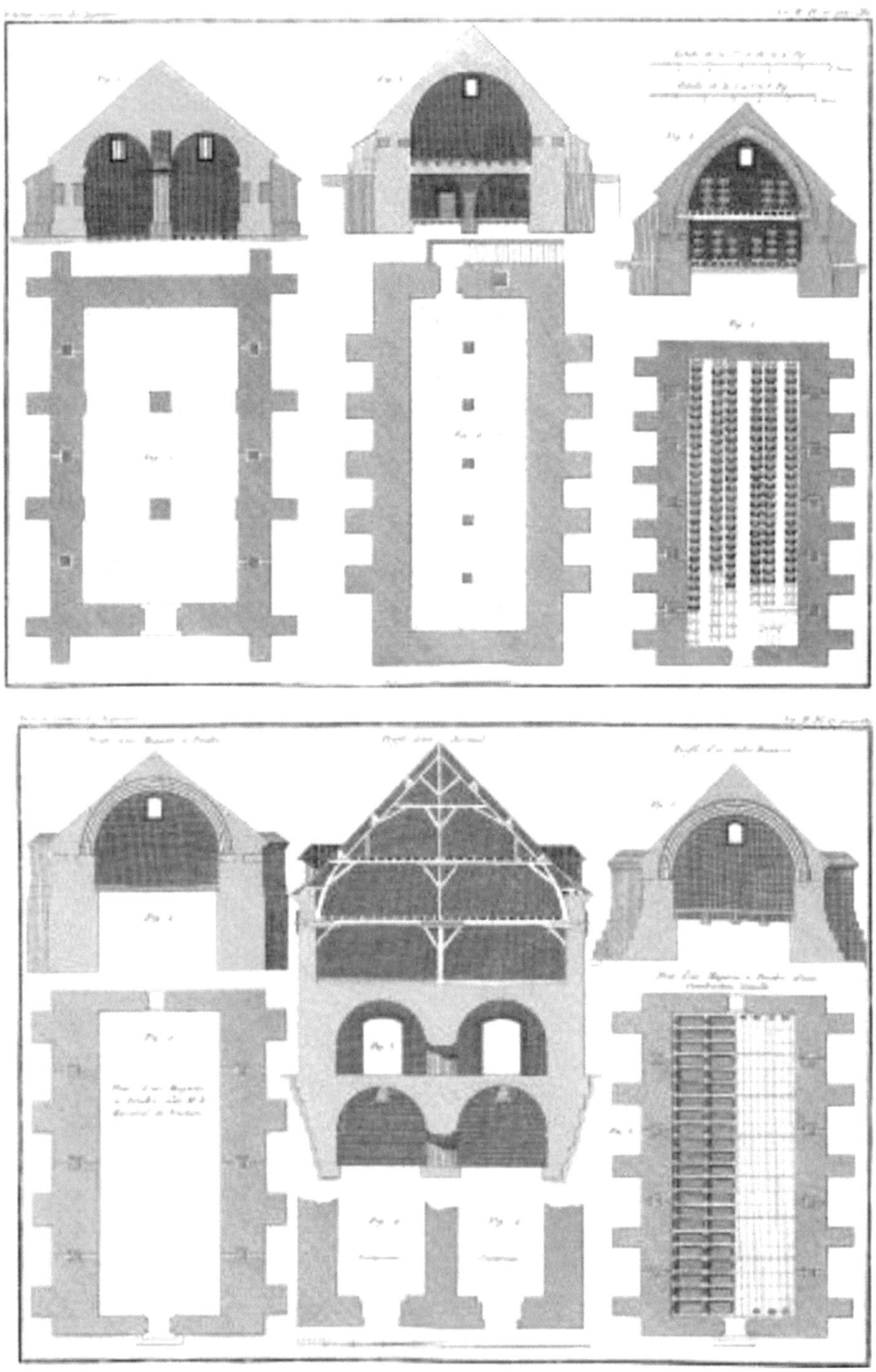

Figure 2.
La science des ingénieurs (1729) of Bernard Forest de Bélidor.

Stereotomy by Abraham Bosse (1604–1676), in addition to *La pratique du trait à preuve de M. des Argues Lyonnois pour la coupe des pierres en Architecture* (1643), in the library of the Barcelona Academy [25].

Another reference work for engineers is *Compendio Mathematico* (1707–1715) by Tomás Vicente Tosca (1651–1723). Treatise XVI in volume V, which is divided into six parts, deals with military architecture (1712). It is available in the academy's library, in Verboom's library (+1744), in Cermeño's library (+ 1790), and in Hermosilla's library (+1776). [26] This treatise deals with arch and vault dimensioning, as well as their collapse mechanisms. It claims that the perfect shape of an arch is a mixed one, made up by the intrados of a round arch and the extrados of a pointed arch. Furthermore, Vicente Tosca improves the three-centered arch or basket-handle arch (*arco apaynelado* or *arco carpanel*), since in Book II, Prop. III, he establishes for the first time the geometrical construction of ovals, which are defined by the length of their two main axes [27].

Yet another work of reference in the libraries of the military engineers was *Traité des Ponts* (1716) by Hubert Gautier (1660–1737) in the libraries of Aylmer (+1788), Aedo Espinosa (+1787), Roncali (+1794), and Cermeño (+1790) [28].The second edition (issued in 1723) includes the famous statement "Ut pondera libra, sic aedificia architectura," referring to the difference in thrust between a round arch (which tilts the balance) and a pointed arch [29]. This edition also includes an additional dissertation: *Augmenté d'une Dissertation sur les Culées, Piles, Voussoirs, et Poussées des Ponts*. The 1728 edition includes a revised and enlarged version of *Dissertation sur l'Epaisseur des Culées des Ponts, sur la Largeur des Piles, sur la Portée des Voussoirs, sur l'Erfort & la Pesanteur des Arches à differens surbaissemens*. [30].

The *Oeuvres de Monsieur Maroitte* (1717), by Edme Mariotte (1620–1684), was also used for teaching purposes. The most interesting part is Volume 2, which includes *Traité du mouvement des eaux et des autres corps fluides, divisé en V parties*; this work is in Verboom's library (+1744) and in Cermeño's library (+1790) [31], previously published by de La Hire (1686).

However, the main references for the Spanish military engineers were definetly the works of Bernard Forest of Bélidor (1698–1761): the *Nouveau cours de Mathématique* (1725), *La science des ingénieurs* (1729), and the first (1737) and second (1739) volumes of his *Architecture hydraulique* (1739), in the Library of Verboom (+1744), Espinosa (+1787), Burgo (+1788), Cermeño (+1790), and Juan Miguel de Roncali (+1794). In *Nouveau cours de Mathématique* (1725), Bernard Forest de Bélidor discusses a practical application of masonry mechanics to the construction of gunpowder magazines [32]. Bélidor calculates the abutment for a barrel vault and for a third-point arch. He includes a table summarizing the size of the pieds droits depending on their curvature and their location, specifying as well if they are supporting the basement floor or the roof. In *La science des ingénieurs dans la conduite des travaux de fortification et architecture civile* (1729) (**Figure 2**), Book II, Chap. III. Prop V, Bélidor establishes the curvature that a vault should have so that all its parts have the same weight and are well balanced (the result is a curve with the shape of a catenary). Bélidor differentiates five vault topologies in military constructions: barrel vaults, third-point vaults, surbased vaults with an elliptical profile, flat vaults, and vaults with a shape that results from the chain [33].

2.1 Gunpowder warehouses in the military architecture treatises

Military architecture treatises of José Cassani (1673–1750), developed at the end of the seventeenth and eighteenth centuries, make reference to the construction of these warehouses, especially if they have an element of high resistance, such as having been made bombproof or having been constructed underground (whether in manmade excavations or in caves) [34].

The principal work of reference is *Maniere de fortifier selon la methode de Monsieur de Vauban*, of Sébastien Le Prestre Vauban (1633–1707), edited by the abbot Du Fay in 1681. The morphology of the gunpowder warehouses with a double enclosure is defined in that treatise, together with the design of its roofing [35].

In the Spanish treatise *El Ingeniero Primera Parte, de la Moderna Architectura Militar* (1687), by Sebastián Fernández de Medrano (1646–1705), that question is

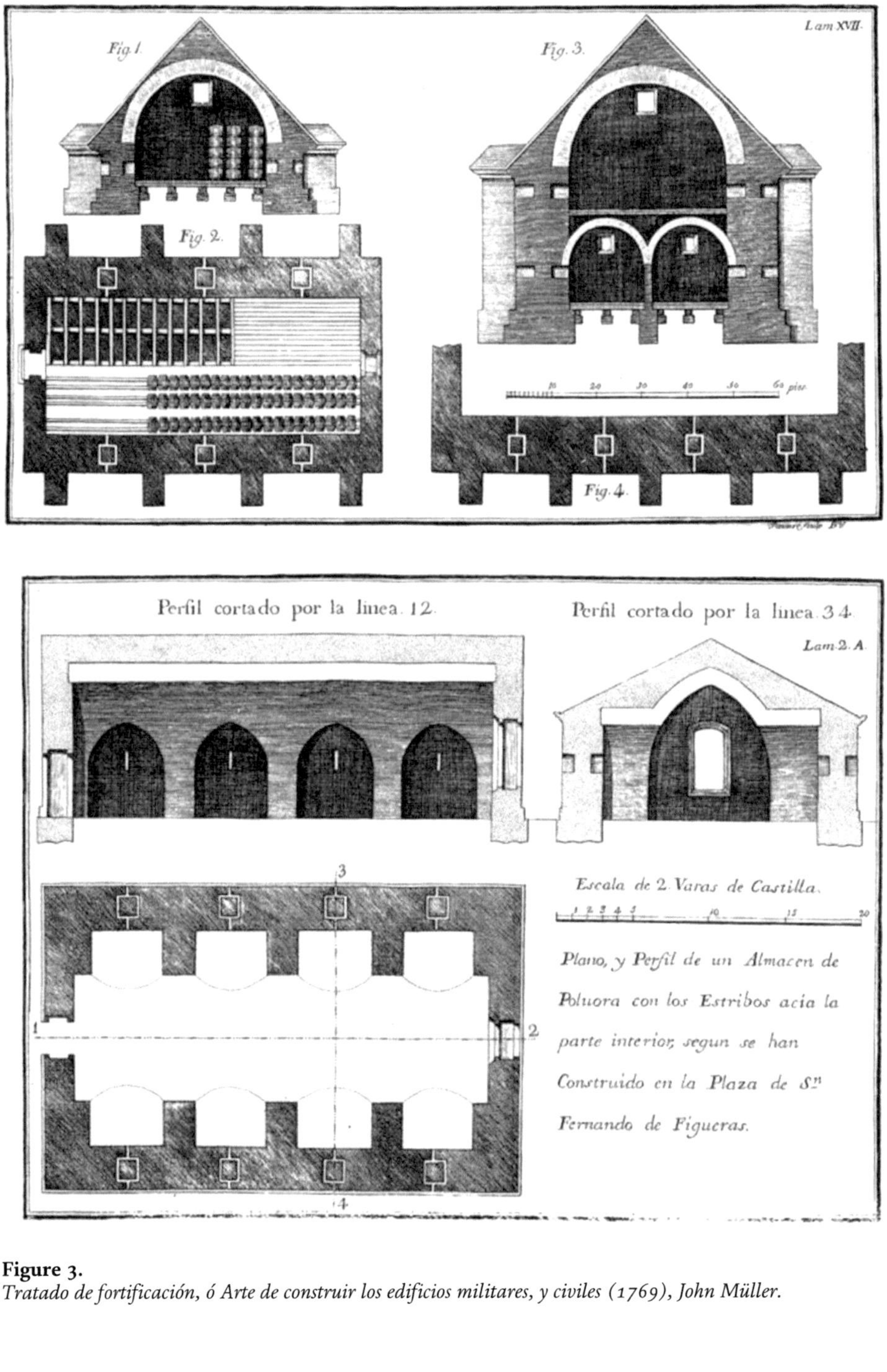

Figure 3.
Tratado de fortificación, ó Arte de construir los edificios militares, y civiles (1769), John Müller.

solved with a masonry vault of 4 feet (0.324 m) width of Spanish "vara" [36]. Subsequently, in the *El Architecto Perfecto en el Arte Militar* (1700), the thickness is increased from 12 to even 14 feet [37]. In the *Escuela de Palas ò sea Curso Mathematico* (1693), attributed to José Chafrion (1653–1698) in the *Libro de Arte Militar*, XI Treatise [38], constructions similar to the Vauban's constructions exist. This vault is solid and is built with lime masonry, which differs from Sebastián Fernández's vault.

In *Nouveau cours de Mathématique* (1725), Bernard Forest de Bélidor considered a practical application of masonry mechanics for the construction of gunpowder warehouses. He determined the abutment for a canon vault and for a tiers-point pointed arch. In a table, he synthesized the dimension of the pieds droits, in ratio with its curve and localization [32]. The importance of these strategic elements obligates Bélidor to dedicate the entirety of Chapter IX of the Book IV of *La science des ingénieurs dans la conduite des travaux de fortification et architecture civile* (1729) (**Figure 2**) to the construction of gunpowder warehouses. In Book II, Chap. III, Proposition V, Bélidor presented that the curve must be given to a vault in order not only to keep up its weight throughout but also to keep itself in equilibrium [33]. As a result, its curve would have the shape of a catenary. Thus, for military constructions, Bélidor determined up to five different types of vaults: semicircular, third-tip pointed, elliptical (drawn as a segmental arch), plane, and (the derived forms of the) catenary.

The work of Bélidor is translated into English by John Müller (1699–1784) and published under the title *A Treatise Containing the Elementary Part of Fortification, Regular and Irregular. For the use of the Royal Academy of Artillery at Woolwich* (1755) in Part. III, Sect. XIX, of this translation is entitled *Of Powder-Magazines* [39]. This text is also translated into Spanish by the Mathematics Academy Professor Miguel Sánchez Taramas in Barcelona (1769) under the title *Tratado de fortificación, ó Arte de construir los edificios militares, y civiles* (1729) (**Figure 3**) for the use of the pupils of the mathematics school. For the construction of gunpowder warehouses, their references are Vauban (1681) and Bélidor texts [40].

Élémens de Fortification (1739) by Guillaume Le Blond (1704–1781) is another influential text. With regard to the construction of gunpowder magazines, Le Blond raises Problem II: *Tracer de rempart et le parapet* [41]. Lastly, we would like to highlight *Principios de Fortificación* (1772), a treatise by Pedro de Lucuze (1692–1779) for the academy. With regard to the construction model, in Chapter XIX (entitled *Edificios Principales*), he defines the main characteristics that a building must have *robustness*, *convenience*, and *symmetry*. He draws a distinction between two types of structures according to their robustness: a simple model and a bombproof model. Bombproof structures need a sufficiently thick vault built with stone or brick or a cushioning system consisting of a wooden framework covered with earth [42].

3. Morphologies of gunpowder magazine projects (1715–1798)

Gunpowder magazine projects made by the Spanish military engineers of the eighteenth century are based on previous military architecture treatises. Therefore, most of these ancillary constructions are built bombproof by shielding the roof. A distinction is made between two types of designs: vaults and wooden structures. The latter are protected by elastic components capable of cushioning the impact of a pyroballistic weapon. From a morphological point of view, gunpowder magazines can be classified depending on their protecting enclosures. From a formal point of view, a distinction is made between the following three morphologies:

Gunpowder magazines having a simple construction body, with the outer wall directly exposed to hostile fire. This is the case of the gunpowder magazines built in Zaragoza (1729) [MPD, 39, 041] (**Figure 4**), Cádiz (1749) [MPD, 56, 029], San Sebastián (1738) [MPD, 27, 092], or Peñíscola (1739) [MPD, 18, 262].

Gunpowder magazines having an outer protection enclosure and a simple enclosure for storing the gunpowder. This is the most common type of magazine in the military treatises. Examples that stand out are the gunpowder magazines in Cardona (1718) [MPD, 19, 028] (**Figure 5**), San Sebastián (1722) [MPD, 28, 034], Ceuta (1724) [MPD, 39, 083], Málaga (1724) [MPD, 59, 046], and Gerona (1738) [MPD, 01, 018].

Lastly, gunpowder magazines having an outer protection enclosure and a two-element central construction body (where the inner wall is at ground level and the main magazine is above ground level). This is the case of the gunpowder magazines built in Tortosa (1721) [MPD, 64, 019], Málaga (1721) [MPD, 59, 044] (**Figure 6**), Barcelona (1726) [MPD, 10, 060], Cádiz (1728) [MPD, 08, 236], or Zaragoza (1729) [MPD, 28, 010].

3.1 Wood beam structure supported by load-bearing walls

These gunpowder magazines are built on load-bearing walls which are parallel to the vault's longitudinal axis, with a perpendicular framework of wooden beams. Where the width is small (until about three toises), the project is built with a single span. Examples that stand out are the gunpowder magazines in Tortosa (1721) [MPD, 64, 019] (**Figure 7**), Málaga (1721) [MPD, 59, 044], Barcelona (1726) [MPD, 10, 060], Cádiz (1728) [MPD, 08, 236], or Zaragoza (1729) [MPD, 28, 010]. Where the width is greater, there are two structural spans. The latter projects are divided into three types: firstly, projects where the central body has load-bearing walls and the two vaulted spans are connected by small doors, such as in Cádiz (1728) [MPD, 08, 237] and Cartagena (1745) [MPD, 18, 258] or Alicante (1750)

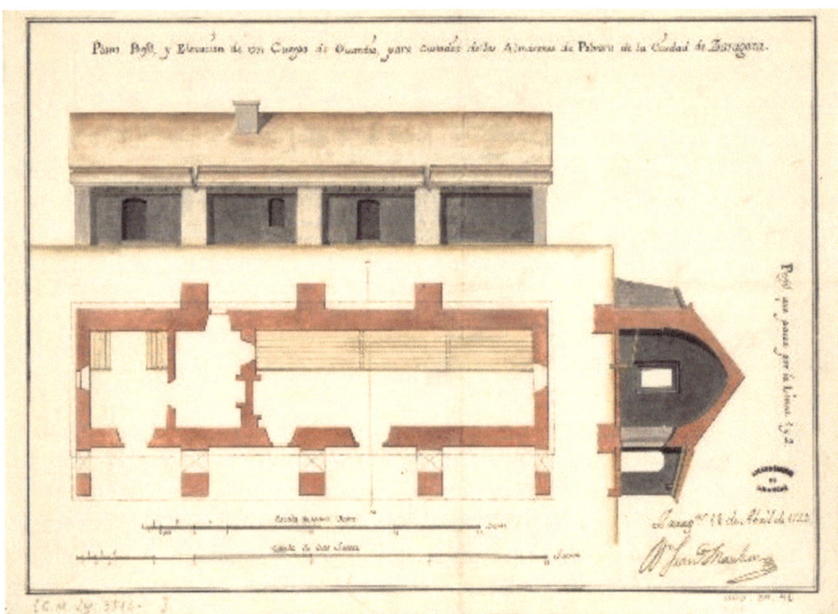

Figure 4.
Zaragoza (1729) [MPD, 39, 041].

[MPD, 65, 088 and MPD, 65, 092]; secondly, projects where the two spans are separated by pillars and the roof is supported by wooden main beams, such as the gunpowder magazines in Hondarribia (1733) [MPD, 65, 044], Cartagena (1745) [MPD, 18, 257], and Tortosa (1798); and thirdly, projects where the roof is

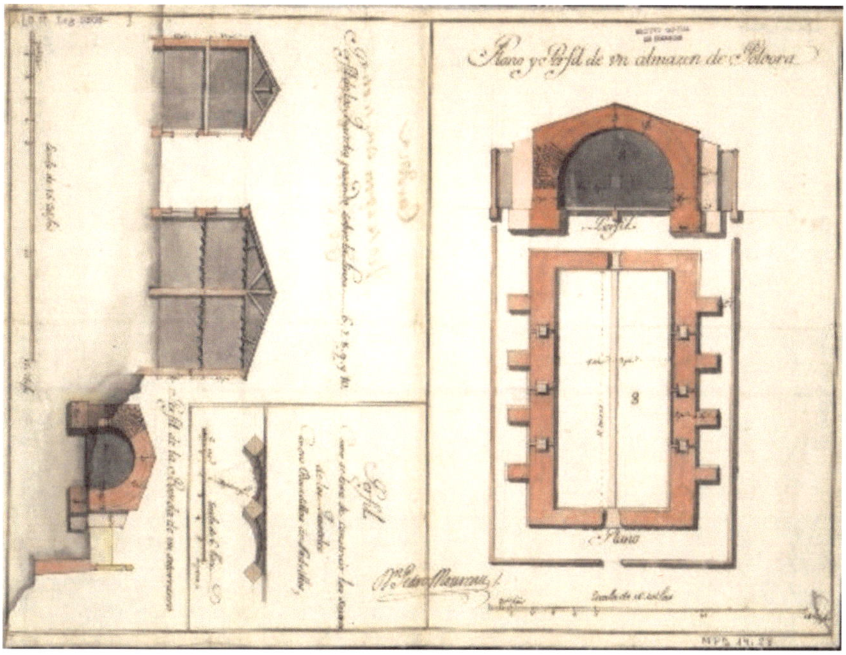

Figure 5.
Cardona (1718) [MPD, 19, 028].

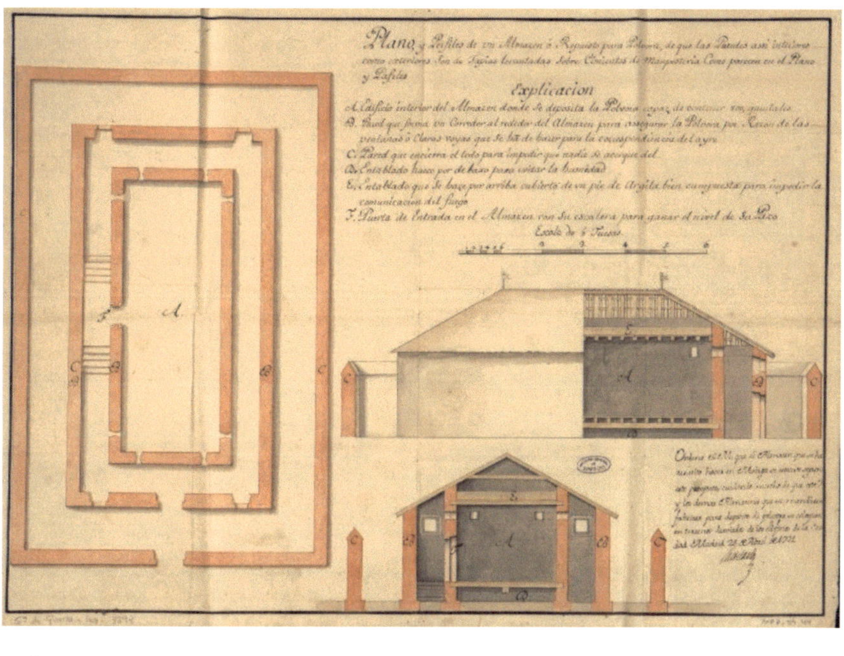

Figure 6.
Málaga (1721) [MPD, 59, 044].

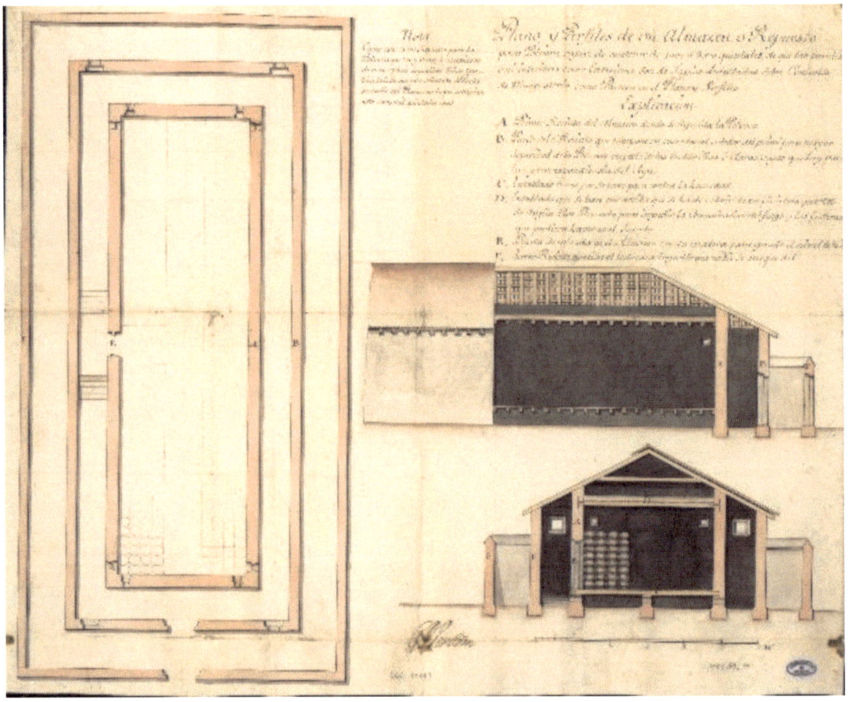

Figure 7.
Tortosa (1721) [MPD, 64, 019].

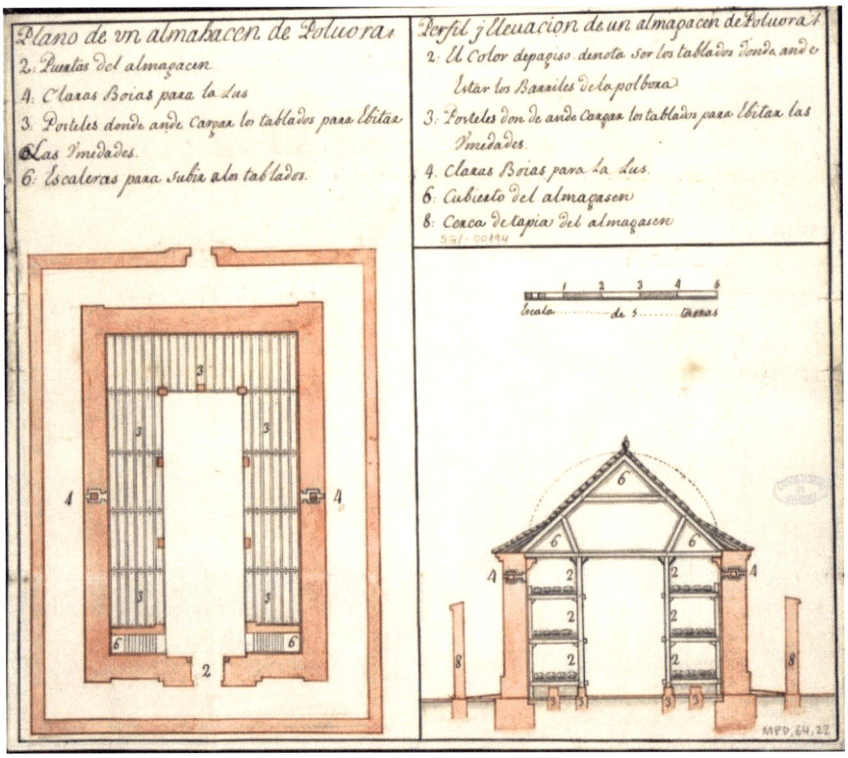

Figure 8.
Málaga (1721) [MPD, 64, 022].

supported by masonry arches, such as in the gunpowder magazines of Ceuta (1735) [MPD, 07, 179] or Lleida (1739) [MPD, 07, 001].

3.2 Wood structure supported by diaphragm arches

Military architecture treatises require the outer walls of the gunpowder magazines to be reinforced with buttresses. Thus, buttresses were used very often as the abutment of diaphragm arches. These types of design emerged later, and structures thus generated may have a single span. This is the case of the gunpowder magazines in Benimàmet (1751) [MPD, 06, 169] (**Figure 4**), Valencia (1756) [MPD, 07, 028] (where the arch abutment is built outwards), and A Coruña (1774) [MPD, 28, 027] (where the abutment is concealed in the interior space, as Müller set out in his treatise (1769)). Other larger powder magazines feature two parallel vaulted spans, for instance, the one in Barcelona (1761) [MPD, 20, 031], with a central pillar between each pair of arches.

3.3 Wood trusses

These are gunpowder magazines having a pitched roof and a timber framing. In some instances a joggle-truss is used, i.e., in the projects for Cádiz (1718) [MPD, 64, 020] and Gerona (1755) [MPD, 10, 073]. In other instances a collar-beam truss is used, i.e., in Barcelona (1731) [MPD, 18, 100 and MPD, 18, 101]. But mainly these magazines are built using the Spanish double-framed roof. This is the case of the gunpowder magazines in Zaragoza (1729) [MPD, 28, 009], Zamora (1734) [MPD, 65, 042], El Ferrol (1772) [MPD, 04, 089], or Cartagena (1795) [MPD, 46, 051 and MPD, 46, 052]. Sometimes they are Spanish double roofs with small variations affecting the inclined tie beams, i.e., in Pamplona (1723) [MPD, 64, 023]

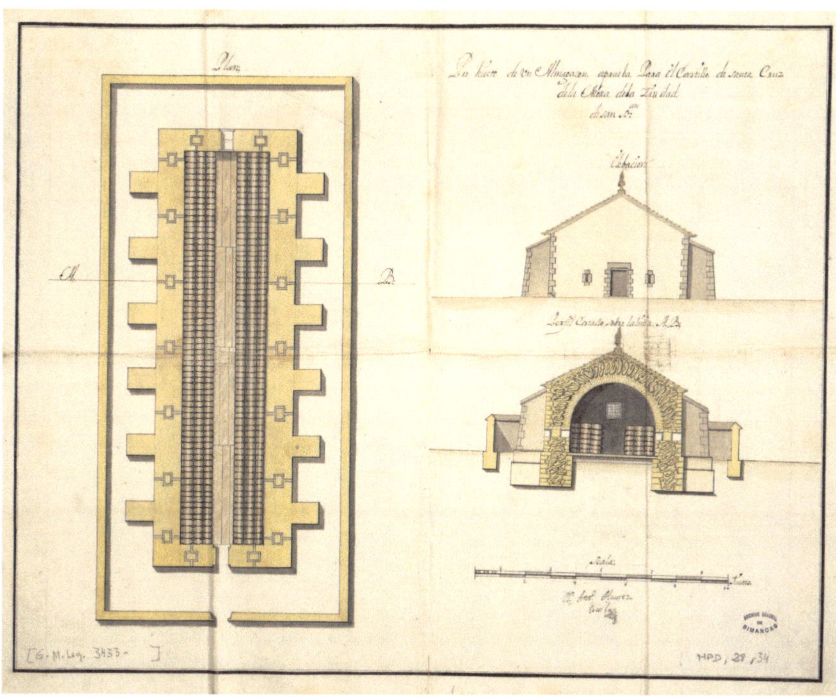

Figure 9.
San Sebastián (1722) [MPD, 28, 034].

and El Ferrol (1738) [MPD, 47, 094]. Other projects feature scissors trusses, i.e., in Monzón (1740) [MPD, 54, 049], Palma (1748) [MPD, 65, 047], and Valencia (1754) [MPD, 06, 170]. The project for Barcelona (1796) [MPD, 46, 035] features a double-framed roof but with two horizontal rafters as tie beams. In other cases, such as the project for Málaga (1721) [MPD, 64, 022] (**Figure 8**), there is a mansard truss with two horizontal rafters (this structure is more similar to the models used in the upper body of military barracks).

3.4 Barrel vaults

Masonry barrel vaults are the most commonly used type of bombproof vaults. The gunpowder magazines are the direct result of military architecture treatises. Because it is such a common construction technique, most of the engineers do not even mention the material used for the vault of the magazine. This is what happened in the projects for Ceuta (1724) [MPD, 39, 083], Longone (1725) [MPD, 12, 222], A Coruña (1738) [MPD, 17, 058], San Sebastián (1738) [MPD, 27, 092], Peñíscola (1739) [MPD, 18, 262 and MPD, 18, 263], Játiva (1748) [MPD, 54, 012], Palma (1748) [MPD, 65, 048], San Sebastián (1750) [MPD, 27, 093], Viveiro (1778) [MPD, 19, 241], and Santa Cruz de Tenerife (1792) [MPD, 05, 033]. With regard to the projects which do specify the material used for the construction of the barrel vault, a distinction must be made between those which use stone for the first layer of the vault and those which use ceramic bricks.

The main projects which use stone masonry include those for Barcelona (1715) [MPD, 18, 097] and Dénia (1748) [MPD, 65, 085]. There is a variant which uses an ashlar arch with wooden beams on top, such as the project for Santa Cruz de Tenerife (1758) [MPD, 18, 050] (this variant was already used by Fernández de Medrano in the previous century). In other cases, ashlar masonry is substituted by

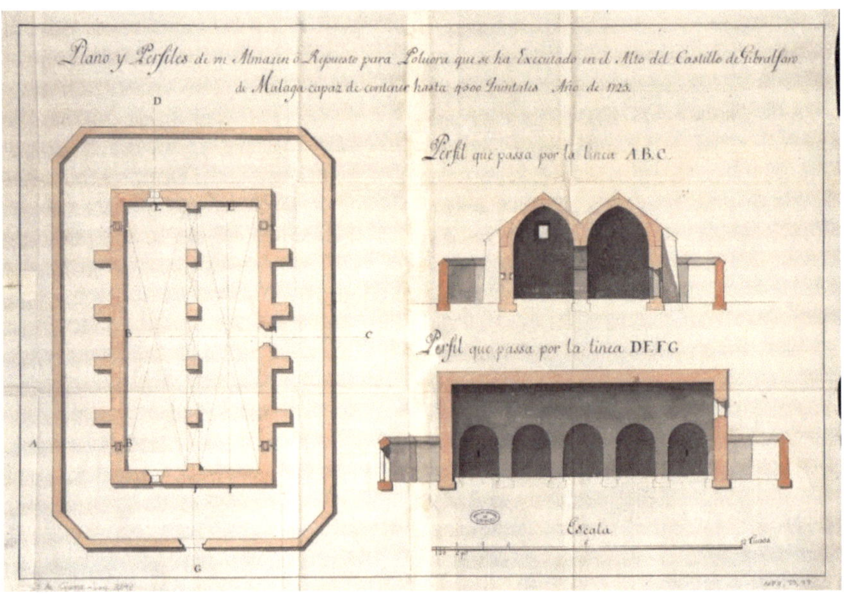

Figure 10.
Gibralfaro Castle in Málaga (1724).

brick masonry with a lime and pebble covering, as in the case of San Sebastián (1722) [MPD, 28, 034] (**Figure 9**).

As for the projects which use ceramic bricks, a distinction must be made between four different types: firstly, those which feature a single-layer ceramic vault, like the ones for Zamora (1738) [MPD, 13, 113] or Ciudad Rodrigo (1739) [MPD, 12, 154]; secondly, those which feature a bottom layer of ceramic bricks, an intermediate layer of stone, and a top layer of lime and pebble, like the gunpowder magazine in Cardona (1718) [MPD, 18, 028]; thirdly, vaults designed with a double layer of ceramic bricks and a top fill of lime and pebble, like in the projects for Pamplona (1718) [MPD, 31, 032; MPD, 31, 033; MPD, 31, 034; MPD, 31, 035; MPD, 31, 036]; and lastly, projects which feature three layers of ceramic bricks, for instance, in Longone, Italy (1728) [MPD, 12, 221], or even four layers, like in Badajoz (1749) [MPD, 65, 045].

3.5 Pointed vaults

Pointed vault structures were described by Bélidor (1729), and Müller (1769) said they are less resistant to bomb impacts than barrel vaults. They belong to the Gothic building tradition, and they need a smaller abutment than barrel vaults, even though magazine walls in military treatises are dimensioned depending on the impact of pyroballistic weapons (those employing gunpowder) and not on the basis of masonry mechanic criteria. This is the case of the gunpowder magazines in Zaragoza (1729) [MPD, 39, 041] and San Fernando de Cádiz (1749) [MPD, 56, 029], which have only one enclosure. It is the same construction type as the projects for Gerona (1738) [MPD, 01, 018] and San Sebastián (1749) [MPD, 27, 094], but these two magazines are protected by an encircling wall, and therefore they have two enclosures. The gunpowder magazine projects for Gibralfaro Castle in Málaga (1724) [MPD, 59, 046 and MPD, 59, 047] (**Figure 10**) and Ceuta (1737) [MPD, 07, 180] feature two parallel vaulted vans which are separated by square pillars supporting round arches, thus forming the central valley of the roof.

4. The curve of equilibrium and the Spanish military engineers

In the construction of gunpowder warehouses, barrel and pointed vaults are generally used, although there are some examples with elliptical vaults, such as that one built in 1694 by Hércules Torelli in Pamplona. This construction was remodelled by Francisco Larrando de Mauleón (1718) [MPD, 31,031] (**Figure 11**) [43]. Mauleón was professor at the Mathematics School of Barcelona and Zaragoza and authored the *Estoque de la Guerra y Arte Militar*, published in Barcelona (1699). The viceroy had ordered the repair of the fortifications and the gunpowder warehouses to make them bombproof. The elliptical vault was replaced by a barrel vault to make it less visible and vulnerable to enemy artillery. It was reinforced and reduced in height according to the concept introduced in the military treatises by General Ambroise (d. 1587) in *Le Timon du Capitaine* (1587) [44].

The simple vault of the warehouse of Montjuïc mountain in Barcelona (1731) [MPD, 07, 057], a project attributed to Miguel Marín (**Figure 12**), is not generated through an arch of circumference. The geometric study reveals that the vault has a length of 16 feet in toise, a rise of 11.5 feet, a width of 3 feet, and a buttress of 7 feet. A geometric element having the shape of a catenary can be drawn running through

Figure 11.
Warehouse of Pamplona, Francisco Larrando de Mauleón (1718) [MPD, 31,031].

the springing points and the key of the vault (**Figure 13**). The shape of this element, which has the same span and rise as the vault, is very similar to the shape drawn in the project.

Other projects, such as the project from Miguel Marín for Tortosa (1733) [MPD, 13, 035] (**Figure 12**), have a span of 21 feet in toise, a rise of 14 feet, a width of 3.5 feet, and a buttress of 7 feet (**Figure 13**). Another similar project is the simple warehouse layout by Juan de la Feriére y Valentín in A Coruña (1736) [MPD, 17, 057] (**Figure 12**), which contains a span of 22 feet in toise, a rise of 14 feet, a width of 3 feet, and a buttress of 7 feet (**Figure 13**).

The design of the pointed vault is initially compared with the catenary, as obtained with a chain over a reproduction of the plans on a larger scale (**Figure 14**). Thus, the arch described by the chain is very similar but not coincidental to the profile of the vault because there are small deviations near the springline of the vault. This deviation is because it is not possible to lay out the catenary with traditional drawing tools, such as rulers and compasses.

The assessment of the original section drawing of the warehouse reveals three compass marks. One point is made over the vertical axis of symmetry of the figure, while the other two are made over the perpendicular axis, slightly below the springline of the vault. An oval was drawn on each project using these compass marks, and the obtained curves were coincident with the curves of the projects. Thus, to draw the projects of the warehouses, both Miguel Marín and Juan de Feriére y Valentín used the geometrical solution of an oval. Therefore, the curves drawn in the three projects are oval, but the major axes of the oval are higher than

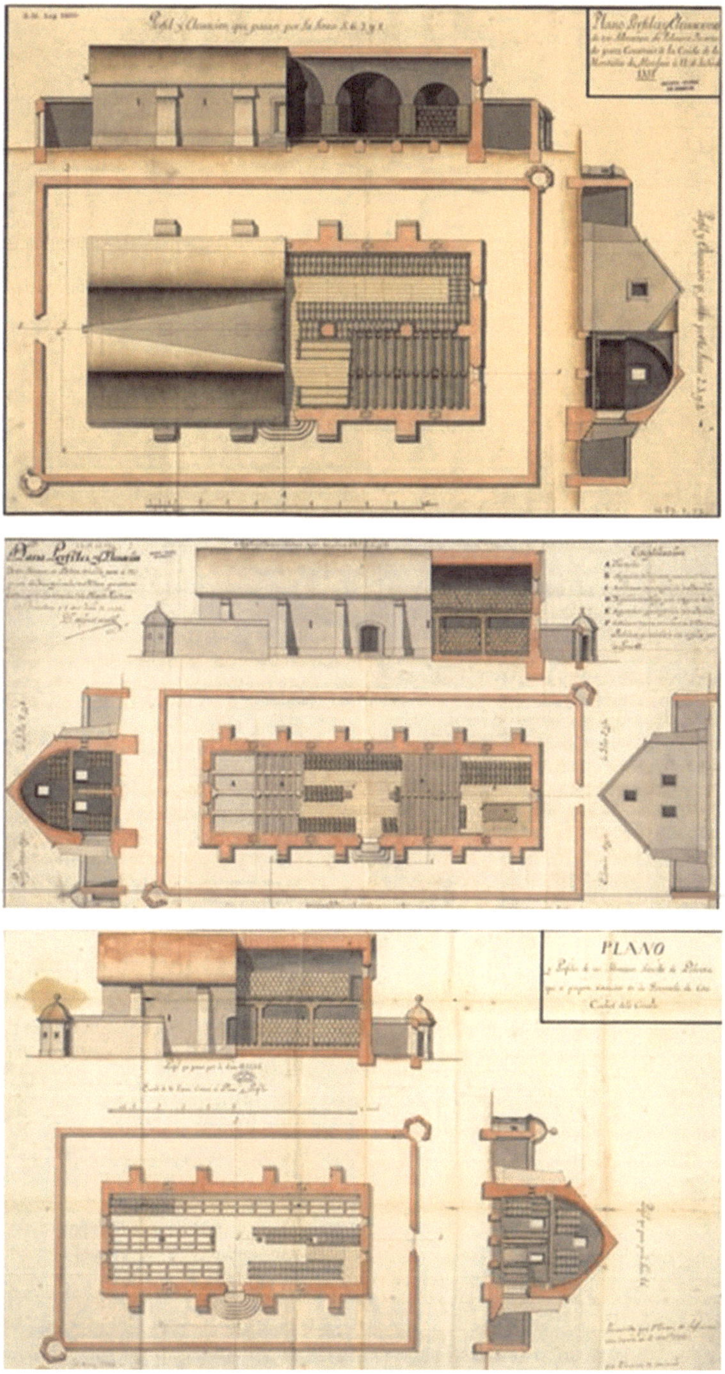

Figure 12.
Gunpowder warehouses: Marín (1731), Marín (1733), and Feriére y Valentín (1736).

the springline of the arch. As a consequence, the curves are not tangential at the springing. So, the military engineers drawn the curve of the vault as an arch *apaynelado, carpanel* of Tosca (1712), or *anse de panier* of Bélidor (1729) (**Figure 15**).

The geometric layout of these vaults, based on ovals, was well known by the eighteenth-century military engineers. They began from the essential feature that oval vaults are tangential to the springline of these building elements. When

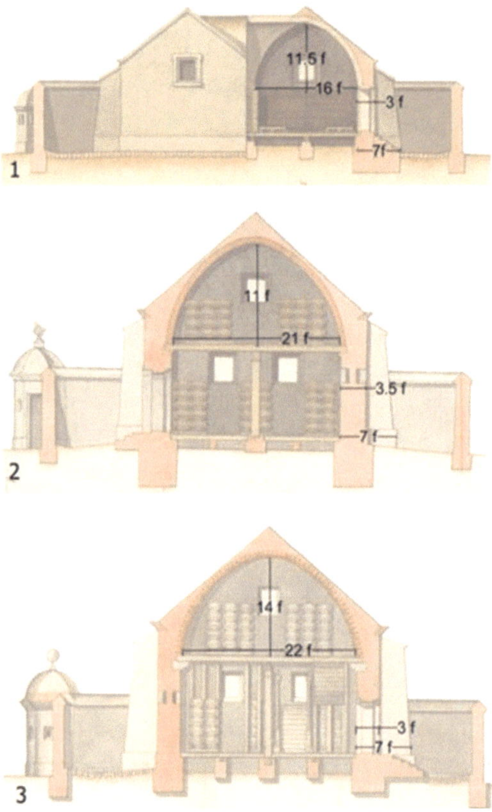

Figure 13.
Metrology of the projects for the gunpowder magazines.

the springline is higher than the axis, a non-tangential curve is obtained, which is a feature of the catenary definitions given by Frézier (1738). Concurrently, Bélidor (1729) specifies the method to lay out the true shape of the catenary vault. By knowing the rise and the span of the vault, the architectonic shape is determined with a hanging chain. Thus, a scale model can be built and can easily be taken to the construction site. By contrast, the layout of the catenary in military engineers' projects is more complex, because it needs the use of an approximation of the catenary through the geometrical shape of a lowered oval.

The ovals are derived from the centres of the circumferences (the compass center points on the paper). They are referred to (O1) for Barcelona (1731) [MPD, 07, 057], (O2) for Tortosa (1733) [MPD, 13, 035], and (O3) for A Coruña (1736) [MPD, 17, 057]. They are consistent with three different types of ovals, and they all share the common feature that the origin of the vertical tangent to the minor axis of the oval is located 1 foot below the impost line (**Figure 16**).

The main ovals' geometric data are shown in **Table 1**, where:

- e1 describes the clear span.
- e2 describes the rise.
- a1 describes the distance between centres of the minor axis.

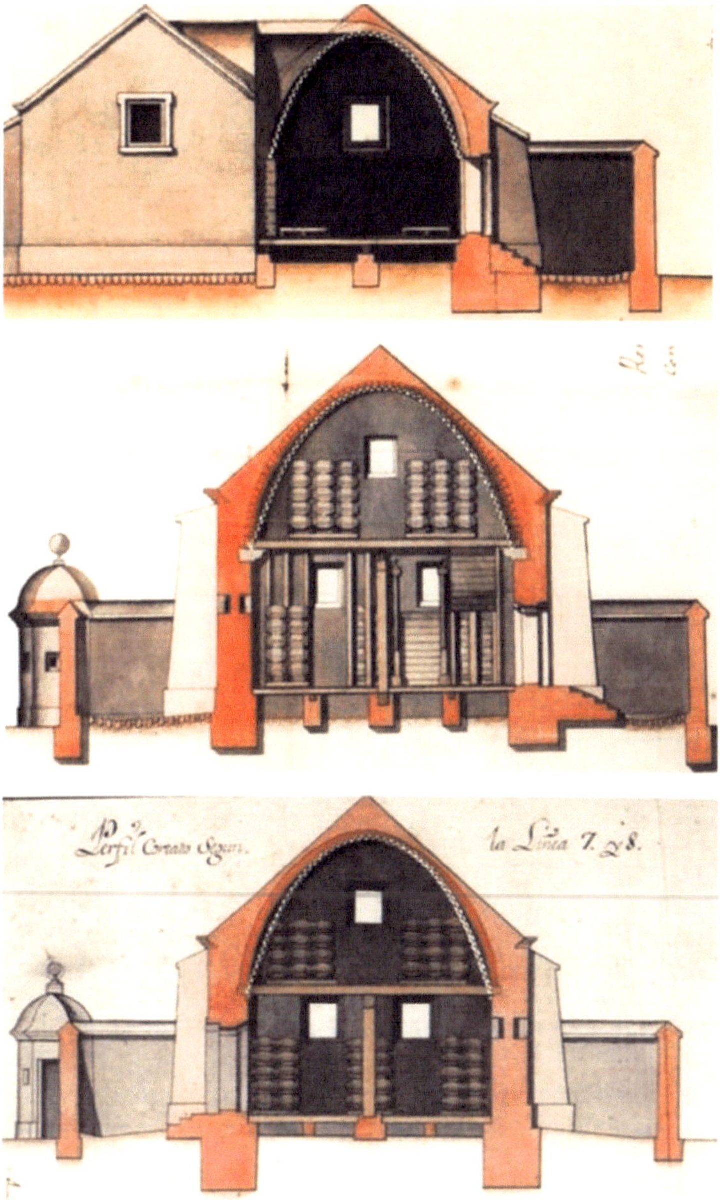

Figure 14.
Chaînette method applied to the projects for the gunpowder magazines.

- a2 describes the distance between the centre of the minor arc and the minor axis.
- d1 describes the ratio between the length of the semimajor axis and the vertical distance from the semiminor axis to the springline (feet).
- d2 describes the ratio between the length of the minor axis and the distance from the center of the major arc to the point of tangency between the major arc and the minor axis.

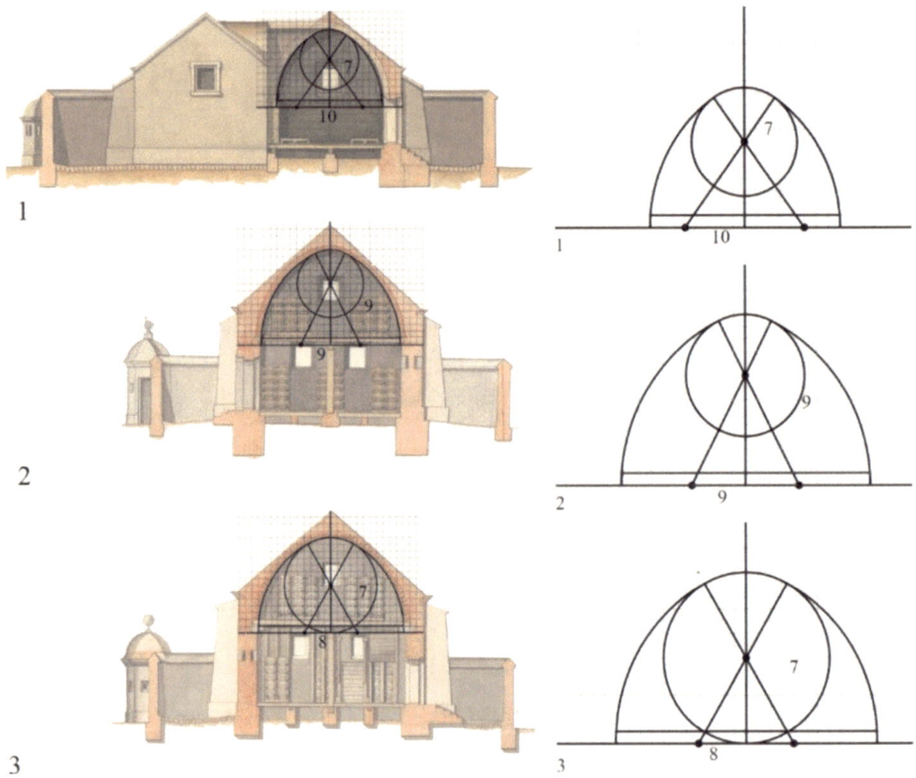

Figure 15.
Oval method applied to the projects for the gunpowder magazines.

- d3 describes the ratio between the length of the semimajor axis and the radius of the oval's minor arc.
- p1 describes the ratio between the semimajor axis and the minor axis.
- p2 describes the ratio between the center-to-center distance on the minor axis and the distance from the center of the semimajor axis to the minor axis.

The ovals used in the layout of the gunpowder magazines are thus used as a reference for purposes of comparison with the cellar's layout. The layout of [O1, O2, O3] is based on a ratio between d3 and e2 of [0.39:0.36:0.50].

In addition, the layout of each oval is compared with a catenary that has the same rise and span, which is drawn using InnerSoft software. According to the results, the inner surface defined between the corresponding geometric shape and the springline is different (1.33 m^2 vs. 0.98 m^2). Furthermore, the ratio between the maximum distance between geometric shapes and the arch's span ranges from 2.14 to 3.44%. From these data, we can conclude that the approximation made by the engineers by drawing ovals in the three projects considered is sufficiently precise for the drawing scale used, between E: 1:90 and E: 1:70. Finally, the curves are compared with an elipse with the same rise and span. The obtained shape is clearly not coincident with the curves of the projects.

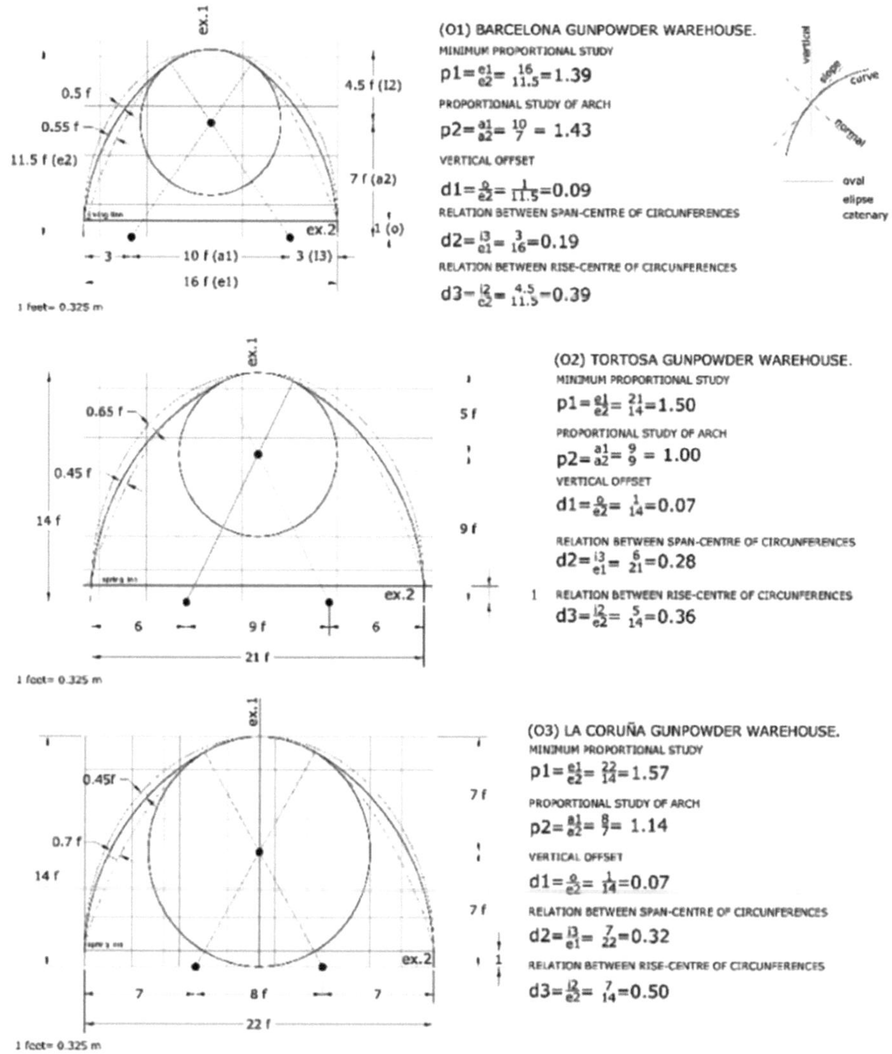

Figure 16.
Geometrical analysis of the vaults of the projects for the gunpowder magazines.

5. The curve of equilibrium and the civil constructions of the eighteenth century

After the Bourbon dynasty's ascendancy to the Spanish throne (1700), Catholic diplomatic and military families of Irish and Scottish origin emigrated under royal protection, preserving their status. O'Connor family was installed in Benicarló in the eighteenth century, and they associated with the McDonnells in the wine export business.

The Bourbon dynasty, which established itself in Spain in 1700 with King Philip V (1683–1746), created the Army Corps of Engineers by the Royal Decree of 17 April 1711. Several Irish families moved to eastern Spain in the mid-seventeenth century. Patrick White Limerick, a trader in agricultural products and wine, and the

O'Connor family settled in Benicarló in 1749. Gaspar White and the O'Gorman family were based in Alicante, whereas the Lilikells, the Tuppers, and Henry O'Shea lived in Valencia. Against this backdrop, the O'Connor family built a Carlón wine cellar in Benicarló in 1757 [45] (**Figure 17**).

The wine cellar's construction, using diaphragm arches, is very similar to the gunpowder magazine projects built by the Army Corps of Engineers in a neighbouring geographical area. These include the project by Carlos Beranger [MPD, 06, 169] for Benimàmet (1751) and the one by Juan Bautista French (1756) for Peñíscola [MPD, 07, 208]. In these projects, as opposed to the O'Connor cellar, the arch abutment is on the outside of the building. Nonetheless, there is a subsequent project by Antonio López Sopeña [MPD, 28, 027] for A Coruña (1774), in which he uses diaphragm arches with inside abutments similars to those used in the Benicarló's wine cellar.

The O'Connors family Benicarló's building was built in 1757. The geometric study of the cellar arches is based on the topographical survey conducted with a laser scanner. The Carlón wine cellar has a rectangular floor plan; its inside measures are 12.42 m in width and 43.01 m in length. In the nave, there are eight diaphragm arches, each having a single two-piece offset-jointed ring and an average depth of 0.60 m. The arches are made of solid ceramic bricks (measuring $0.37 \times 0.18 \times 0.04$ m), and they rest on a limestone base that was brought from Santa Magdalena de Polpís. The top of this base determines the springline of the ceramic arch. The arch's abutment and the outside walls are made of ordinary uneven masonry. The formal characteristics of the arches are different: arch a_1 has a clear span of $e1_{a1} = 9.65$ m and a rise of $e2_{a1} = 5.82$ m, whereas the other seven arches can be grouped together. Their span is within a range of $e1_{a(2-8)} = [9.76–9.69$ m$]$, similar to arch a_1, but their rise significantly differs from the first arch, within a range of $e2_{a(2-8)} = [5.46–5.45$ m$]$. All of the arches share a special feature: they do not have a vertical tangent on the stone base. The angle of incidence (α) of these arches with respect to both vertical sides, left (α_a)and right (α_b), has the following values: $\alpha_{a.a1} = 13.94°$ and $\alpha_{b.a1} = 8.97°$ in arch a_1 and $\alpha_{a.a(2-8)} = [4.49°–1.80°]$ and $\alpha_{b.a(2-8)} = [7.38°–2.58°]$ in arches $a_{(2-8)}$ (**Figure 8**). By statistically analysing the

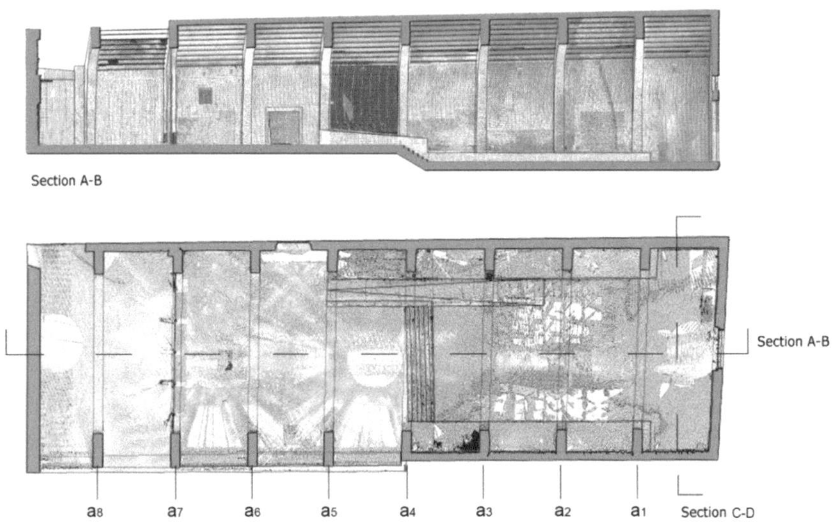

Figure 17.
Floor plan and cross section of the O'Connor cellar in Benicarló (1757).

parameters, for arches $a_{(2-8)}$ the average span calculated is $e1_{a(2-8)}$ = 9.72 m, and the average rise is $e2_{a(2-8)}$ = 5.46 m (**Figure 18**).

It seems that the measurement units used for the construction of the cellar were the toise (194.90 cm) and the toise foot (32.48 cm). Arches $a_{(1-8)}$ have an average span of 29.92 toise feet (9.71 m), with an error of 0.02 m for 30 feet (5 toises, 9.74 m). The rise of arch a_1 is 17.92 feet (5.82 m), i.e., there is an error of 0.02 m in 18 feet (5.84 m), which are 3 toises (**Figure 19**). Arches $a_{(2-8)}$ have a rise of 16.80 toise feet (5.46 m), i.e., there is an error of 0.06 m in 17 feet (5.52 m). The arches are 0.60 m in width (1 + 10/12 feet). Regarding the outside measurements, the nave is 41.50 feet (13.48 m) wide and 92.36 feet (30 m) long, and the arches' abutments are structures measuring 5 + 9/12 feet. The inside length of the cellar is 43.01 m, i.e., 132 + 5/12 toise feet. The enclosure wall on the façade is 2 feet thick; thus, the span-to-arch ratio is 5.75/30 feet (**Figure 20**).

A metrological analysis of the arches in Benicarló's cellar reveals that the eight arches show the same metric relations, i.e., a 5 toise span and a 3 toise rise. The dimension of the catenary arch a_1 are 30 × 18 feet (exactly 5 × 3 toises). The dimensions of the elliptical or oval-shaped arches $a_{(2-8)}$ are 30 × 17 feet. If we follow the hypothesis that the minor axis (either the ellipse minor axis or the oval minor axis) is 1 foot below the impost, then the geometric relation of arches $a_{(2-8)}$ is also 30 × 18 toise feet.

A statistical analysis is now performed on each of the eight arches $a_{(1-8)}$ to determine the difference between the shape of the arches built and the shapes of reference: ellipse (E), catenary (C), and ovals [O1, O2, O3]. The following values are calculated for 29 points on each cellar arch:

a. The average and maximum deviation of these 29 points

b. The angle of incidence on the springline

The mean deviation has a spread of only 0.03 m, which is approximately 0.31% of the arch's span, making it difficult to conclude whether it is a catenary or an oval. The determining feature is that arch a_1 has an angle of incidence on the springline

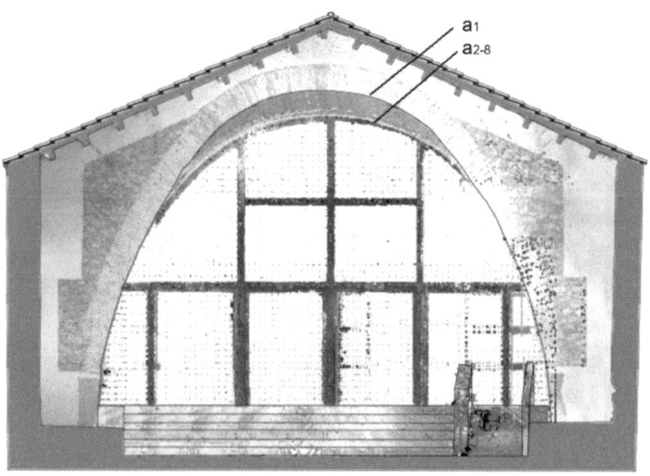

Figure 18.
Transversal section of the O'Connor cellar in Benicarló (1757).

[$\alpha_{a.a1}$ = 13.94°, $\alpha_{b.a1}$ = 8.97°]. Because the catenary's angle of incidence is 19.38°, the geometric shape that most closely resembles the arch is the catenary.

Conversely, the statistical analysis of the remaining seven arches $a_{(2–8)}$ shows that the geometric shape that they most resemble is the ellipse, with an average deviation ranging between 0.001 and 0.015 m. The range for oval-shaped arches is 0.006–0.186 m (*arco apaynelado* or *arco carpanel* according to Tosca, i.e., three-centred arch or basket-handle arch; or *anse de panier* according to Bélidor), so the

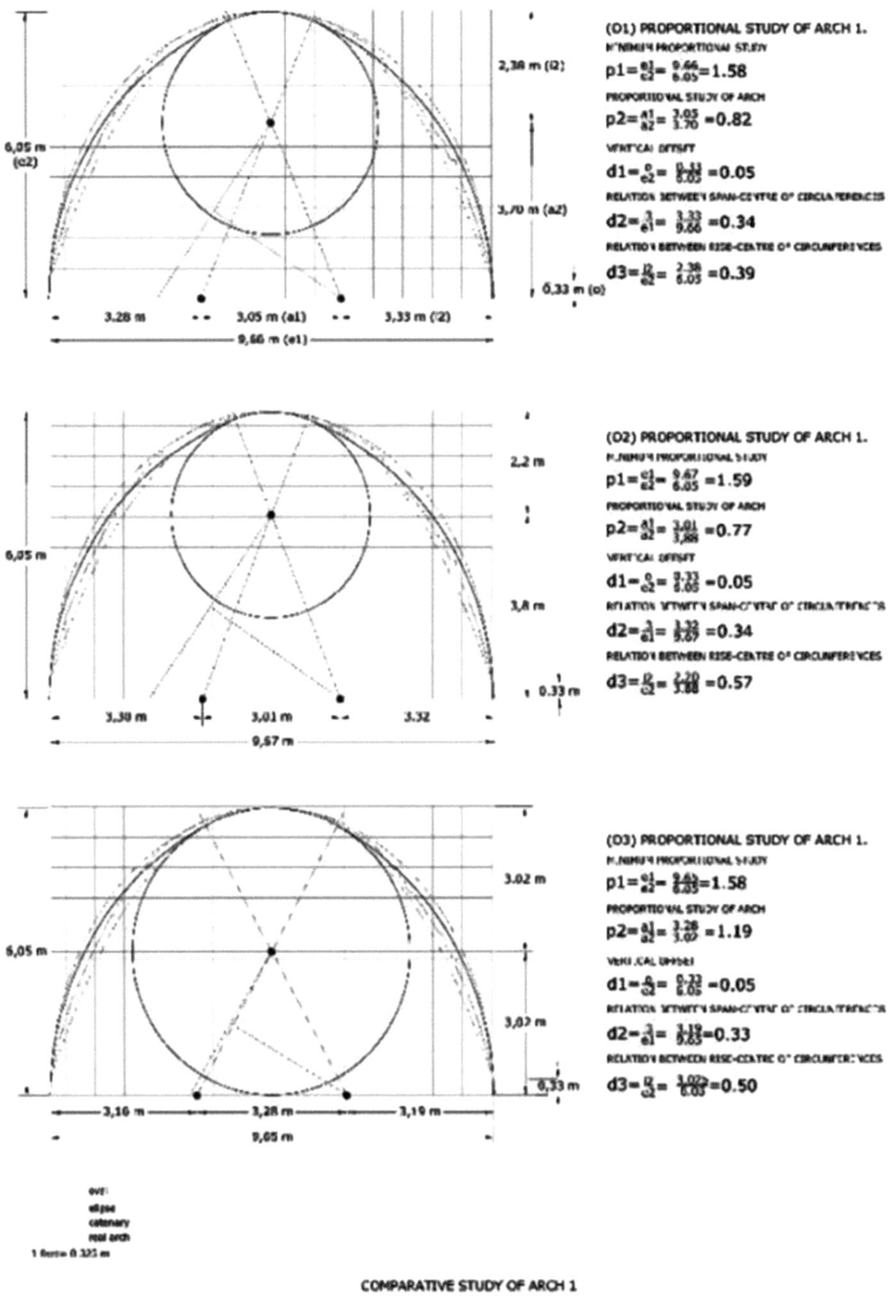

Figure 19.
Geometrical analysis of the arch 1 in the O'Connor cellar (1757).

curves used by the Bourbon engineers to layout the projects for the gunpowder magazines are very similar to the cellar's arches. Nevertheless, the angle of incidence on the springline tends not to have a vertical tangent, which is a fundamental feature of both the ellipse and the oval in the arches considered here. In the springline of these arches, the angle of incidence ranges between $\alpha_{a\ .a5}$ = 1.54° and $\alpha_{b\ .a5}$ = 7.38°.

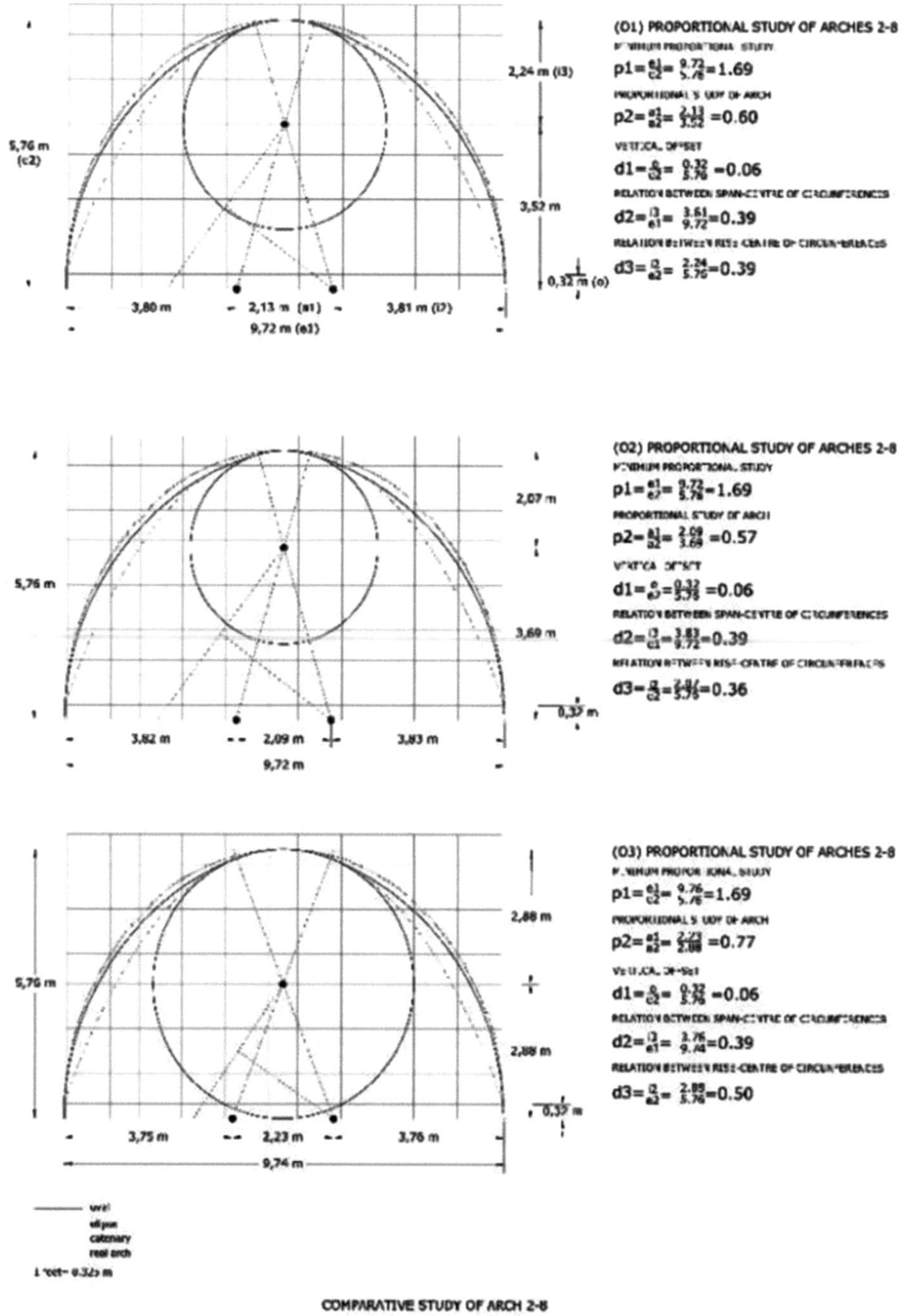

Figure 20.
Geometrical analysis of the arches 2–8 in the O'Connor cellar (1757).

Thus, arch a_1 resembles a catenary arch, whereas the other seven arches $a_{2–8}$ tend to be ellipses. These seven arches do not have a vertical tangent on the springline because their horizontal axis has been moved 1 foot below the arch's springline. As defined by Frézier (1738), the shape of the catenary has the following essential property: the vertical line which is tangent to the curve at the springline does not form a right angle with the horizontal plane. Therefore, geometrically, the catenary can be understood as any curve that does not have a vertical tangent at its springline. This is what happens in the springline of St. Paul's dome in London [11], which was designed by Christopher Wren in collaboration with Robert Hooke [46]. Otherwise, it should be noted that from a mechanical perspective, catenary arches are an optimal solution to build masonry arches, since the material has very low tensile strength.

Finally, from the construction point of view, the catenary shape can be approximated using other geometric forms such as ovals or ellipses, under the condition that there is not a vertical tangent at the springline. The catenary shape forms a barycentric axis, which minimizes the tensions on a linear element that is subject to only vertical loads. In the arch, the inverted catenary shape prevents the appearance of stresses other than compression stresses.

Thus, there are two hypotheses regarding the construction of the wine cellar. The first one is that the construction work was started from the inside toward the façade; thus, arches $a_{(2–8)}$ were constructed before the catenary arch a_1. The second hypothesis is that the construction work began with arch a_1. According to the second hypothesis, there is also a difference between both types of arches: on the first brick courses from the springline of arch a_1 (the first 9 courses on 1 side and the

Figure 21.
Springing of arches no. 1 and no. 2 in the O'Connor cellar.

first 17 courses on the other side), the ring is 0.36 m wide. On the remaining seven arches, the ring is 0.60 m wide (just like the arch's depth). It is clear that less ceramic material is necessary for the construction of arch a_1 than for the other seven elliptical arches $a_{(2–8)}$ (**Figure 21**).

6. Conclusion. The origin of the catenary arch in Spain

The assessment of some drawings of gunpowder warehouses, found in the collection of *Mapas planos y Dibujos* (MPD) of the General Archive of Simancas (*Archivo General de Simancas*, AGS) (AGS 2014), has revealed the use of the chain theory in Miguel Marín's projects for Barcelona (1731) and Tortosa (1733) and Juan de la Feriére ones in A Coruña (1736). A built evidence has also been found: the Carlón wine cellars in Benicarló, built by the O'Connors family from Ireland (1757). The analysis of these examples proved the theory of the chain arrival to Spain during the first half of the eighteenth century. However, 50 years before Antoni Gaudí, Catholic families emigrating from Scotland and Ireland already initiated some of the catenary's form mathematical theory in some practical uses, a theory that begun to be taught at the Mathematics Academy of Barcelona in 1720.

This paper addresses the introduction of the concept of the catenary arch in Spain before the nineteenth century. After an exhaustive review of the theoretical framework, some cases are assessed. The aim of the research is to find out if the mechanical concept of the chain was used by the Spanish military engineers and by the exiled English engineers, who built several wine cellars in Spain. Thus, we intend to determine whether there is any geometrical relationship between the layout of several gunpowder magazines made by Spanish military engineers in the 1730s and the construction of a civil building—the Carlón wine cellar in Benicarló (1757)—in which catenary arches may have been used.

The assessments of the gunpowder warehouses by Miguel Marín for Barcelona (1731) and Tortosa (1733) and by Juan de la Feriére y Valentín in A Coruña (1736) are only a mere 4.05% of the projects analysed. However, they prove the intention to lay out the vault as a catenary. These authors knew that in a catenary the tensility in the shape of a hanging chain has the same compression values in the inverted geometrical figure. These engineers had a vast knowledge of the mechanical principles of the modern theory for masonry. From a scientific perspective, catenary vaults are the most interesting because they introduce the principles established by Hooke (1676). Both the arches of gunpowder magazines and the arches $a_{(2–8)}$ of Benicarló were laid out using the geometrical construction of an oval. Otherwise, the location of the horizontal axis of the ovals under the springline reveals the application of one of the characteristics of the catenary. This causes that the vertical line which is tangent to the curve in the springing does not form a right angle with the horizontal, so they used the chain's theory in the layout of the projects.

Formally, if the distance between the axes and the springline of the arch is small, then the angle of incidence has a minimum influence on the thrust and the line of pressure. Otherwise, the location of the axis under the springline reveals the intention to minimize stresses in this point and in the neighbouring areas, even though the final mechanical influence is small.

Although there is no evidence of the construction of the gunpowder warehouses, it is possible to confirm the use of catenary arches in the construction of the Carlón cellars of the O'Connor in Benicarló (1757). There are significant differences between the measures of the arches of the gunpowder magazines (maximum span: 22 feet; maximum rise: 14 feet) and the arches of the Benicarló cellar (span: 30 feet; rise: between 17 and 18 "toise" feet, until the springline). In addition, the span-to-rise ratio of the oval arches in the gunpowder magazine studies is [1.39:1.57], whereas in Benicarló, this ratio is [1.67:1.76]. It can be concluded that arch a1 is a catenary arch, whereas arches $a_{(2–8)}$ tend to be elliptical. Arches $a_{(2–8)}$ show the special feature that their (x) axis is located below the springline; therefore, the tangent of the curve on the springline does not form a right angle with the horizontal. This is a feature of the definition of the catenary.

The theory of the equilibrium curve, followed by most of the British engineers, became known to the Bourbon military engineers through the academy of mathematics in the eighteenth century, and it was used by some immigrants of English origin, such as the O'Connors, a century before the modernist architecture of Antoni Gaudí.

Acknowledgements

The author thanks the members of the PatriARQ group research: PhD Agustí Costa Jover, PhD Sergio Coll Pla, and Arch. Mónica López Piquer.

Author details

Josep Lluis i Ginovart
School of Architecture Barcelona, Universitat Internacional de Catalunya, Barcelona, Spain

*Address all correspondence to: jlluis@uic.es

References

[1] Huerta S. Structural design in the work of Gaudí. Architectural Science Review. 20004;**49**:324-339

[2] Llorens JI. Wine cathedrals: Making the most of masonry. Proceedings of the Institution of Civil Engineers: Construction Materials. 2013;**166**: 329-342

[3] Martinell C. Gaudí i la Sagrada Familia comentada per ell mateix. Barcelona: Ayamà; 1951. p. 29

[4] Lacuesta R. Gaudi a través de Cèsar Martinell, la relación personal entre Antoni Gaudí i Cèsar Martinell. In: VV AA, editor. Els arquitectes de Gaudí Barcelona: Col·legi Oficial d'Arquitectes de Catalunya. 2002. p. 120

[5] Alsina C, Serrano G. Gaudí, Geometricamente. Gaceta de la Real Sociedad Matemática Española. 2002;**5**: 523-558

[6] Domenech J. La fábrica del ladrillo en la construcción catalana. Anuario de la Asociación de Arquitecto de Cataluña. 1900. pp. 37-48

[7] Rondelet JB. Traité théorique et pratique de l'art de bâtir, V2. Paris: Chez l'auteur, en clos du Panthéon. 1804. pp. 137-145

[8] Millintong J. Elementos de Arquitectura, escritos en ingles por John Millington,V2. Madrid: Imprenta Real; 1848. pp. 472-477

[9] Graus R, Martín-Nieva H. The beauty of a beam: The continuity of Joan Torra's beam of equal strength in the work of his disciples-Guastavino. Gaudi and Jujol International Journal of Architectural Heritage. 2013;**9**:341-351

[10] Hooke R. A Description of Helioscopes, and Some Other Instruments. London: Printed by T.R. for John Martyn; 1676. p. 31

[11] Heyman J. Wren, Hooke and Partners. In: Huerta S, editor. Proceedings of the First International Congress on Construction History. Madrid: Instituto Juan de Herrera; 2003. pp. 1-11

[12] Stevin S. Het Eerste Bovck van de Beghinselen der weegcons. In: De Beghinselen der Weeghconst. Beschreven duer Simon Stevin van Brugghe. Leyden: Inde Druckerye van Cristoffel Plantijn, By Françoys van Raphelinghen. 1586. p. 41

[13] Bernoulli J. Analysis problematic antehac propositi de inventions lineae descensus a copore gravi percurrendae uniformiter, sic ut temporibus aequalibus aequales altitudines emetiaatur. Actae Eruditorum. 1690; Mensis Maji: pp. 217-219

[14] Bernoulli J. Solutio problematis funicaularii. Actae Eruditorum. 1691; Mensis Junii: pp. 274-276

[15] Huygens C. Dynastae Zulichemii, solutio problematis funicularii. Actae Eruditorum. 1691; Mensis Junii: pp. 281-282

[16] Gregory D. Catenaria. Philosophical Transactions of the Royal Society. 1697; **19**:637-652

[17] Stirling J. Lineae Tertii Ordinis Neutonianae. Oxonia: Whistler; 1717. pp. 11-14

[18] Poleni G. Memorie istoriche della Gran Cupola del Tempio Vaticano. Padua: Stamperia del Seminario; 1748. pp. 30-50

[19] Heyman J. Poleni's problema. Proceedings of the Institution of Civil Engineers. 1988;**84**:737-759

[20] Belidor B. La science des ingénieurs dans la conduite des travaux de fortification et architecture civile. Paris: Chez Claude Jombert; 1729. pp. LI 43-LI 45

[21] Belidor B. La science des ingénieurs dans la conduite des travaux de fortification et architecture civile. Paris: Chez Claude Jombert; 1729. pp. LII 1-LII65

[22] Couplet P. De la poussée des voûtes. Mémoires de l'Académie Royale des Sciences de Paris1729. pp. 75-81. lams. 4-7

[23] García de la Huerta V. Bibliotheca Militar Española. Por Vicente García de la Huerta. Madrid: Antonio Pérez Soto; 1760. pp. 56-129

[24] Derand F. L'architecture des voûtes ou l'art des traits et coupe des voûtes. Paris: Sebastian Cramoisy; 1643

[25] Bosse A. La practique La pratique du trait à preuve de M. des Argues Lyonnois pour la coupe des pierres en Architecture. Paris: Imprimerie de Pierre des-Hayes; 1643

[26] Tosca V. Compendio mathematico: Que comprehende Arquitectura civil, montea, y canteria, arquitectura militar, pirotechnia, y artilleria. Tomo V. Valencia: Por Antonio Bordazar; 1712. pp. 81-252

[27] Tosca V. Compendio mathematico: Que comprehende Geometria elementar, arithmetica inferior, geometria practica. Tomo I. Valencia: Por Antonio Bordazar; 1707. pp. 292-295

[28] Gautier H. Traité des ponts, ou il est parlé de ceux des romains & ceux des moderns. Paris: Chez André Cailleau; 1716

[29] Ht G. Traité des ponts, ou il est parlé de ceux des romains & ceux des moderns. Augmenté d'une Dissertation sur les Culées, Piles, Voussoirs, et Poussées des Ponts. Paris: Chez André Cailleau; 1723 p. Frontispiece

[30] Gautier H. Traité des ponts, ou il est parlé de ceux des romains & ceux des moderns. Nouvelle edition. Augmenté d'une Dissertation sur les Culées, Piles, Voussoirs, et Poussées des Ponts, avec pluisers Devis et Reglaments faits à ce sujet. Paris: André Cailleau; 1727. pp. 335-412

[31] Mariotte E. Oeuvres de Monsieur Maroitte de l'Académie des Sciences divisées en deux tomes. Leide: Chez Pierre Vander Aa; 1717. pp. 326-476

[32] Bélidor BF. Nouveau cours de Mathématique à l'usage de l'Artillerie et du Génie, où l'on applique les parties les plus útiles de cette science à la théorie et à la pratique de différents sujets qui peuvent avoir rapport à la guerre. Paris: Chez Charles-Antoine Jombert; 1725. pp. 490-498

[33] Bélidor BF. La science des ingénieurs dans la conduite des travaux de fortification et architecture civile. Paris: Chez Claude Jombert; 1729. pp. V2 1-V265

[34] Cassani J. Escuela Militar de Fortificación ofensiva y defensiva. Arte de fuegos y de esquadronar. Madrid: Antonio Gonçalvo de Reyes; 1705. p. 86

[35] Fay D. Maniere de fortifier selon la methode de Monsieur de Vauban... Par monsieur l'abbé Du Fay. Paris: Veuve Jean-Baptiste I Coignard & Jean-Baptiste II Coignard; 1681. pp. 171-172

[36] Fernández S. El Ingeniero Primera Parte, de la Moderna Architectura Militar. Dividido en dos Tomos. Brusselas: En casa de Lamberto Marchant; 1687. pp. 238-240

[37] Fernández S. El Architecto Perfecto en el Arte Militar dividido en cinco

Libros. Brusselas: En casa de Lamberto Marchant; 1700. pp. 238-240

[38] Chafrion J. Escuela de Palas ò sea Curso Mathematico. Tomo I. Milan: En la Emprente Real, por Marcos Antonio Pandulpho Malatesta; 1693. pp. 175-177

[39] Muller J. A Treatise Containing the Elementary Part of Fortification, Regular and Irregular. For the Use of the Royal Academy of Artillery at Woolwich. London: Printed for A. Millar; 1755. pp. 215-222

[40] Muller J. Tratado de fortificación, ó Arte de construir los edificios militares, y civiles escrito en ingles por Juan Muller. Barcelona: Por Thomas Piferrer; 1769. pp. 357-370

[41] Le Blond G. Elemens de fortificaction dedie's a son altesse Monseigneur le Prince Charles de Lorraine. Paris: Chez Charles-Antoine Jombert; 1739. p. 46

[42] Lucuze P. Principios de fortificación, que contienen las definiciones de los terminos principales de las obras de Plaza y de Campaña. Barcelona: Thomas Piferrer; 1772. pp. 85-90

[43] Laorden C. Los ingenieros españoles en la creación del arma. Memorial del Arma de Ingenieros. 2005;75:63-103

[44] Bachot A. Le Timon du Capitaine Ab. Bachot, lequel conduira le lecteur parmi les guerrières .... Paris: Au faubourg Saint-Germain-des-Prés; 1587. pp. 20-21

[45] Constante JL. Benicarló 1841–1965; el Tránsito de la Antigua Villa del Maestrazgo hacia la ciudad contemporánea. Tomo III. Benicarló: Onada Edicions; 2012

[46] Addis B. Las Contribuciones de Christopher Wren Y Robert Hooke Al Nacimiento de Ingeniería de la Construcción Moderna. In: Huerta S, López F, editors. Actas del Octavo Congreso Nacional de Historia de la Construcción, 1–11. Madrid: Instituto Juan de Herrera; 2013. pp. 1-11

Chapter 3

Healthcare Military Logistics at Disaster Regions around the World: Insights from Ten Field Hospital Missions over Three Decades

Michael Naor

Abstract

The Israeli Defense Force Medical Corps deployed airborne medical relief operations to disaster regions, inflicted by natural (earthquake, typhoon, and tsunami) and man-made catastrophes. Missions operated around the globe, in Africa, Asia, Caribbean, Europe, and the Middle East. In this study, based on literature review and interviewing of commanders and participants of ten of such missions operating in nine countries (Armenia, Rwanda, Kosovo, Turkey, India, Haiti, Japan, Philippines, and Nepal), we analyze and outline the principals in assembling and operating these missions. Deployment of the relief operations was swift, to address the needs as soon as possible, even at the cost of partial pre-assessment and a wide margin of uncertainty. This was compensated by the design of multi-disciplinarian and self-sufficient and independent units with flexible operative modes, enabling improvisations to cope with unexpected medical and operative needs. The experience gained in these missions led to a well-defined methodology of assembly and deployment of foreign field hospital in a short time. The review shows an evolutionary pattern with improvements implemented from one mission to the other, with special adaptations to address specific requirements and accommodate language, national culture barriers, and ethical dilemmas.

Keywords: disaster relief, field hospital, military logistics, creativity, collaboration

1. Introduction

Global-scale disasters such as the recent earthquake in Haiti, striking without warning, cause massive loss of life, and produce widespread damage to local infrastructure, including medical facilities [1]. Disasters can lead to great loss of life especially if they hit densely populated regions with limited resources and poorly constructed habitation. Furthermore, damage to roads and transportation systems and difficulties in the rescue and evacuation process can impede accessibility to care [2].

Many countries and organizations around the world developed logistic and medical systems designed to cope with disasters. Dispatched foreign field hospital

IntechOpen

(FFH) is one type of medical relief system. The World Health Organization/Pan-American Health Organization defines a field hospital as "a mobile, self-contained, self-sufficient healthcare facility capable of rapid deployment and expansion or contraction to meet emergency requirement for a specified period of time [3]."

The Israeli Defense Force (IDF) Medical Corps developed a model of airborne FFH [4]. This model was structured to deal with disaster settings, requiring self-sufficiency, innovation and flexible operative mode in the setup of large margins of uncertainty regarding the disaster environment. The current study is aimed to critically analyze the experience, gathered in ten such missions deployed in nine countries (Armenia, Rwanda, Kosovo, Turkey, India, Haiti, Japan, Philippines, and Nepal).

The rest of the study is organized as follows. We provide a literature review of healthcare humanitarian aid to disaster areas and a formal definition of a foreign field hospital. The methodology being used is case study. Data was collected by interviews conducted in Israel with senior military staff who actively commanded the humanitarian missions in the disaster areas. Supplemental information was gathered from secondary sources cited in paper. We analyze a series of ten case studies over time period of three decades that provide insights in regards to FFHs deployed by the Israel Defense Forces (IDF) to assist in different types of disasters around the world such as Haiti, Turkey, India, Rwanda, Armenia, and the Philippines. We conclude by sketching future research opportunities that can further develop this field of study.

2. Background

A field hospital is an independent health care facility, which is deployed rapidly for emergency purposes, following the request of the affected country. It is important that delegation and recipient countries clarify in advance the details on responsibilities, chain of command, working protocol with authorities and law enforcement agencies, facilities, installation, and operational process of the FFH in order to avoid any misunderstanding. Both parties need to know the details on the date when the FFH will be operational on site, the FFH equipment and services to be provided, the number of medical staff and their qualifications and experiences, the location of the FFH, and its duration of stay. Next, the components of field hospital deployment are described.

2.1 Mission definition and timing of deployment

Disasters around the world with the potential for the need of international medical assistance are assessed by Israeli governmental bureaus (Ministries of Foreign Affairs, Health, National Security and others), as well as by military offices and local non-governmental organizations. Sending a preliminary assessment team is important. This was the case, for instance, in the pre-FFH era, during the ongoing Cambodian disaster in 1979, were prior assessment of needs, combined with fund raising, led to an incorporation of a drafted team into a Red Cross field hospital in Sakeo, Thailand. More recently, for example, a special assessment team was en route to Haiti 11 h after news of the earthquake reached Israel [5]. An assessing advancing team to Japan evaluated the need for a full scale functional FFH, given the damage to local healthcare system and the medical needs at the disaster zone, coordinated the efforts with the local authorities and regional healthcare providers, defined the required location of the operation, and assessed specific irradiation risks [6].

Acute disaster settings often require immediate assistance, precluding time consuming prior assessment. Furthermore, by the time relief operation arrives, conditions and needs might change substantially, especially if the time required for deployment is extended. Therefore, the Israeli FFH was often one of the first international humanitarian missions active on ground, adopting the principal of "just on time and just in place", at the price of incomplete assessment and a large margin of uncertainty. To compensate for that, the FFH was designed in a way to meet unexpected situations, first by being composed of a multidisciplinary team, and second, by being self-sufficient and independent. Lastly, initiative with numerous improvisations with the help of local agencies and manpower helped coping with unexpected situations. For example, the Armenian mission, operated within a roofed stadium, transformed into a city hospital with the use of plastic sheets stretched on cables, which divided the space into functioning departments. This obviously required a substantial aid provided by local authorities and medical staff [7].

2.2 Location of the FFH

Since swift air deployment is essential to operate expeditiously, missions were airborne, deployed usually in military Hercules airplanes (that enable transportation of vehicles) and occasionally in commercial aircrafts for long-distance missions such as in Haiti/Japan/Nepal. Location of the medical relief operation was usually decided before arrival, and coordinated by pre-assessment team based on dialogs with local health care system and logistic headquarters. Issues taken into account were accessibility to patients, safety (regarding aftershocks in earthquake scenarios, or appropriate safe surroundings in a war zone), and proximity to air fields (for supplies and evacuation in case of emergency). For instance, an intact, roofed municipal sports center provided an adequate shelter and convenient location for the FFH in the snowy Kirovakan. The gymnastic stadium was divided into four functional areas by stretched cables from which black polyethylene sheets were hung, while supplies and surgical rooms were placed on the podium. In Zaire, sleeping quarters were located within an unfinished, fenced, and easily protected private house adjacent to the field hospital. In the missions to Bhuj, India and to Port-Au-Prince, Haiti, soccer fields were chosen as the operation site, because it's a well confined area, usually with one/two entrances, has walls (protection), and its size is adequate [8, 9].

2.3 Communication systems

In a chaotic post-disaster environment, there is a need to utilize both long-range systems to communicate with the delegation's country of origin, and short-range systems to enable communication between site of field hospital home base and local authorities, ambulances, helicopters, as well as delegations from other countries deployed in disaster area. It includes standard walkie-talkie (130–170 MHz), loudspeakers, telephony, fax, internet, email and video conference. Range, spectrum of radio frequencies, bandwidth, weight, size, ease of usage, reliability, batteries life, and cost are important factors in determining which systems the delegation should bring to disaster area. Caution should be taken when patrolling in disaster area with long antenna near collapsed wiring in an earthquake setting. Standard military VHF radio (30–75 MHz) that are non-dependent on local network, proved to be useful in IDF missions for communication with neighboring military units from various countries.

In IDF mission to Rwanda, military VHF systems were utilized for communication with vehicles moving at the range of up to 30 km from the headquarters at the field hospital, which also covered mobile short-distance communication between the hospital and the sleeping quarters. In 1999, at Adapazri, Turkey, short wave communication (telephone and Internet) relied on a high frequency (HF) radio transceiver in the range of 3–30 MHz; in 2010 at Port-Au-Prince, Haiti, a direct satellite channel was established with an 8 GB bandwidth; and in 2011, at Minanisanriku, Japan, broadband global area satellite internet network (BGAN inmarsat) enabled Wi-Fi communication [6].

2.4 Electronic medical data storage and handling

A computerized hospital administration information system has capability to gather rapidly information, analyze it, and present it to medical team. It can also give pharmacist in charge data control over release of medical supplies and provide alerts regarding need to replenish developing shortages of critical items. The IDF designed and used in Haiti such an information system which included: identification and demographic information, photo album, admission notes and status, survey of injuries by body system, laboratory and imaging studies, surgical reports, diagnoses, and discharge summary [10]. The usage of such an electronic medical record in mega-disaster scenario ensures medical accuracy, and lowers risk of losing information in chaotic environment when patients are transferred between FFHs from different countries, or when delegation returns back to home country and give control over FFH facility to local healthcare authorities as occurred in most missions. For instance, in Haiti, bar-code readers were used to facilitate patient's registration upon entry to a specific department within FFH and to minimize manual data entry errors. The database of passport-like photographs was useful for family members to locate their relatives and it was suggested for future designing customized radio frequency identification (RFID) technology in order to track patients in disaster area [11]. Such technologies are developed in Israel as part of a national system for disseminating information on victims during mass casualty events [12].

3. Methods

This article synthesizes ten medical relief operations in disaster settings, carried out during the last three decades in the form of deployed FFH, in a particular pattern designed and executed by the IDF Medical Corps. We interviewed over period of four years (2011–2015) physicians who actively participated in the IDF disaster relief missions from 1988 until 2015, as chief medical officers and other personnel who have vast experience in the logistics, policy, and health ministry aspects involved in this humanitarian domain from their service in the Israel Home Front Command. Several of them were also highly ranked in United Nations Disaster Assessment and Coordination [UNDAC] and the Department of Peacekeeping Operations at the United Nations Headquarters in New York. Therefore, they have expert-knowledge about the administrative aspects of collaboration between countries during relief missions. Each interview lasted about 2 h and was recorded and transcribed with permission from key informant. Interview questionnaire guide can be provided as appendix. After conduction exhaustive literature review of principal medical and auxiliary publications, we integrated information detailing the assembly of the missions, (manpower selection and training), their operative modes (supplies and equipment, medical data storage and handling, communication systems), capacity (number of beds, collaboration with other delegations), and termination protocol in order to ensure continuity of care by local medical staff their operative modes and outcome.

A body of knowledge was accumulated over the years by the IDF Medical Corps from deploying numerous relief missions to both natural (earthquake, typhoon, and tsunami), and man-made disasters, occurring at nine countries in different regions of the globe (Africa, Asia, Caribbean, Europe, and Middle East). Longitudinal studies of this sort which juxtapose different humanitarian missions can be helpful in learning and making better decisions in the case of future disasters. Indeed, our study shows an evolutionary pattern with improvements implemented from one mission to the other, with special adaptations to address specific requirements and to accommodate to language and national culture barriers [13].

An important trait of the Israeli FFH pertains to the medical staff selection and training: the staff of the FFH (physicians, nurses, pharmacists, etc.) is recruited in a very selective process [14]. It is composed from mixture of reservists and actual duty soldiers drafted for the voluntary mission. Knowledge of local languages at the disaster area (Russian, French, etc.) is an important criterion for staff selection. Missions to war regions such as Goma, Zaire, were complemented by armed soldiers that also served as stretcher carriers. Additional personnel included laboratory, logistic and communication technicians. For risk assessment two members of Israel national committee for nuclear energy joined the mission to Japan, equipped with dosimeters for continuous monitoring of irradiation [15].

Importantly, the chosen personnel are composed roughly 2/3 of people who participated in past missions and 1/3 new recruits, in order to transfer knowledge gathered between missions and to create an organizational body of experience pertaining to humanitarian aid. This experience has often been enriched by previous practice gained in military medical units in combat regions, unfortunately prevalent in Israel and surrounding countries.

Based on IDF experience at Adapazari, Turkey, it is recommended, a (nurse):(physician) ratio of (1–1.5):(1), as opposed to a (2.5–3):1 ratio in regular civilian hospitals because paramedics and medics are available for active assistance [16]. These nurses have to be specialized, work longer and more intensive shifts than in a regular hospital. Consequently, physicians need to assist in classic nursing issues.

Adjustments in hospital structure were made during missions. Thus, IDF FFHs functions ranged from primary care and first aid clinics as in the Kosovo and Japan missions, to regional first echelon for patients released from ruins in Turkey, to municipal hospitals, as happened in India or Armenia, and to a medical referral center, as happened in Haiti, Nepal and the Rwandan disasters. In this last example, the operative mode and structure changed over time, in parallel with needs. This mission served initially as a regional cholera camp, but with the recognition of its capabilities, it became a referral center for trauma and other surgical cases, for patients with meningitis and other complicated medical conditions, as well as for critically ill babies requiring intensive care settings.

The functional structure of the FFHs changed accordingly. In Rwanda, a triage and rehydration facility changed into adult and pediatric wards, with a latter addition of expanding departments for surgical/orthopedic/obstetric patients and for those with non-diarrheal critical infections, such as meningitis [17, 18]. The FFH in India was setup in a fully self-sufficient tent encampment. It provided variety of surgical and diagnostic procedures such as: orthopedics (soft tissues, amputations, fracture reduction, external fixation), plastic surgery, skin grafts, debridement/reconstruction, appendectomy, caesarian section, pediatric neonatal intensive care unit, and deliveries. The FFH in Adapazari, Turkey (1999) served for few days as a first level facility for injured population rescued from wreckage, principally providing surgical and orthopedic surgical facilities and managing patients with crush syndrome and associated renal failure [19]. At later stage, beyond the salvageable rescue period, the hospital principally provided first aid and primary care for the nearby population, as the number of patients with acute and chronic medical, pediatric and neonatal conditions exceeded that of traumatic cases [20]. The heterogeneous mixture of medical staff enabled the transformations that took place in the operative mode of the FFH.

In the same way, the Haiti mission coped in the first days with injuries caused directly by the earthquake, with very busy orthopedic and surgical units, doubling the surgical capacity by cross-over mixed teams concomitantly addressing needs for various surgical disciplines. A few days later, when patients with less urgent medical needs arrived, staff assignments, organization of unites, and hospitalization policy were readjusted.

4. Results

Our case studies cover a variety of field hospitals deployed by the IDF. Descriptive information of the missions is described in **Table 1** and statistics is presented in **Table 2**. The studied disasters vary in their type, size, and number of casualties, addressing different types of required medical services.

All earthquakes in our case studies had a magnitude of higher than 7 on the Richter scale, associated with mass casualties and damage to local health facilities, requiring foreign assistance. FFH can confront various levels of acuity. If the number of casualties is extremely high (Haiti), one may expect confronting severely wounded patients. If the damage is mainly to the infrastructure (Nepal), one will confront more chronic conditions. It depends on the number of injured people seeking medical care, number of other FFH, how fast the team arrives, the baseline standard of care, damage to local facilities, etc.

For example, in the missions to Armenia or India, most treated casualties in the FFH were principally survivors with minor or intermediate injuries and patients with a variety of acute and chronic medical conditions seeking substitute for the non-functioning local health systems. In some cases, as in Haiti, the FFH served as a tertiary medical center, in the absence of domestic alternatives until the establishment of an appropriate substitute, in this case the floating hospital USNS Comfort [21, 22]. In the two missions to Turkey, the FFH served as a buffer and regional second echelon, relieving pressure from nearby functional local health systems [23, 24]. In Japan, due to the rather late arrival and efficient local medical and evacuation systems, the team work took the form of a primary care service.

Our study also reviews missions addressing medical needs of displaced and crowded refugees in two other countries: Kosovo, and Rwanda [25]. While the Kosovo mission addressed anticipated ongoing medical needs of such a population, the relief operation to Goma, one of many heterogeneous medical relief missions orchestrated by the UNHCR, faced overwhelming outbreak of lethal diarrheal epidemics that exceeded any reasonable capacity [26]. In addition to treating such patients this mission became a referral center for complicated patients transferred from other health facilities.

The evolution of IDF humanitarian operations started by deploying mobile clinics to disaster areas, the first time in Kefalonia (1953), and subsequently in Skopje (1963), or by joining international relief operations such as a Red Cross hospital for Cambodian refugees in Thailand in 1979 [27]. Later, it developed into the adaptable structure of FFHs where the scale was tailored to the disaster arena. The first full scale FFH was in Armenia and later in Turkey, and subsequent missions.

Country	Description of FFH
Armenia	In December 1988, a 7.1 magnitude earthquake occurred in Kirovakan, Soviet Armenia. The IDF medical corps deployed a field hospital to Kirovakan. The FFH team included general and orthopedic surgeons, anesthesiologists, experts in rehabilitation, internal and emergency medicine, nephrology, and pediatrics. The medical relief operation was originally designed to serve as a pediatric rehabilitation center combined with dialysis facilities, as requested by the Soviet authorities but eventually provided primary care. The majority of patients received ambulatory treatment, but there were additional trauma cases, gynecology-obstetrics, and a few acute general surgical cases. Sources: [7, 18, 17].
Rwanda	In July 1994, IDF deployed 3 teams sequentially for 6 weeks to Goma, Zaire, following a tribal strife in Rwanda with consequently displaced population subjected to large scale epidemics (principally cholera and dysentery) and famine. The length of the operation requiring team substitution every 2 weeks, with replacements and supplies arriving by subsequent cargo airplanes, enabling continuous prolonged operation. In each team there were experts in internal medicine and pediatrics with subspecialties, clinical microbiology/ tropical medicine, critical care, anesthesiology and neonatology, general and orthopedic surgeons, and gynecologists. The FFH comprehensive multi-disciplinary facilities provided primary and secondary care. The FFH composed of a triage unit, pediatric, medical and surgical wards, and diagnostic facilities. Sources: [7, 18, 25, 36].
Kosovo	The conflict in Kosovo in the 90s escalated in 1999, causing more than one million people from Kosovo to flee from their country to the neighboring countries of Macedonia and Albania. In April 1999, the IDF provided medical services to the refugees. The structure of the hospital was composed of several wards: emergency room, internal medicine, obstetrics and gynecology, pediatric and neonatology, delivery, pharmacy, laboratory x-ray, and security. Twenty hours after arriving in Macedonia, the FFH became functional in the Brazda camp. The IDF field hospital became the referral center for all others primary medical teams. Most of the patients were treated for infections (because of poor sanitary conditions in the refugee camps), exhaustion, and chronic illness (heart disease, diabetes, etc.). Sources: [26].
Adapazari, Turkey	On August 17, 1999, a major earthquake (7.4 Richter) occurred in western Turkey. The city of Adapazari was severely hit. The Israeli field hospital was sent by the Israel Defense Force (IDF) command. The IDF field hospital located in Adapazari provided advanced surgical and medical services; it included trauma care and life saving surgeries and was ready to accept patients in 24 h after arrival on site. The site included 5 beds for intensive care treatment and 80 beds for general hospital admission including internal medicine, obstetrics and gynecology, and surgery. The hospital staff was overall composed of 102 personnel acting as a secondary referral center. Sources: [6, 16, 19, 20, 23, 28]
Duzce, Turkey	In Nov 1999, an earthquake of 7.2 magnitude struck Turkey, this time in the region of Duzce. The IDF medical corps Field hospital was sent 3 days after the disaster. It functioned for 9 days, aiming to substitute for a part of the damaged medical facilities. It acted as a secondary referral center providing specialized and surgical care The hospital structure included seven clinical branches: emergency room (triage), operation room (OR), surgical intensive care unit, internal medicine, orthopedics, pediatrics, obstetrics, and gynecology. The Israeli Field hospital managed to fill the gap in the local medical system, and during its peak operation, its capacity was 300 patients per day. The field hospital focus was on secondary medical care rather than primary and urgent care. Sources: [24].
Bhuj, India	On January 26, 2001, a 7.7 Richter earthquake occurred in India, with the epicenter located in the city of Bhuj. The IDF-led relief activity in India departed within 84 h after recruiting personnel from both regular army and reserve units and initiated hospital activity at site on day six. The field hospital had a fully self-sufficient tent enactment with 30 beds and included auxiliary services units such as radiology, laboratory and medical supplies, and a logistical support unit. The total number of personnel deployed for the India operations was 97. Sources: [8].

Country	Description of FFH
Port au Prince, Haiti	A 7.2 Richter magnitude earthquake struck Haiti on January 2010. The Israel Defense Medical Corps Field Hospital was on site and operational 89 h after the earthquake and provided medical care to many patients during its 10 days of operation. The hospital brought all required supplies in order to stay independent and provide fast deployment, including medical requirements such as antibiotics, imaging machines and lab facilities, and energy sources and accommodations. The Field Hospital consisted on 121 hospital staff members, divided in different units, including medical, surgical, pediatric, orthopedic, gynecologic, ambulatory and auxiliary. The capacity of the Field hospital was 60 inpatient beds, which could be expanded to 72. Sources: [5, 6, 9, 10, 21, 22, 35, 42]
Japan	An earthquake of 9.0 on the Richter scale struck Japan on March 11, 2011. It caused a Tsunami that washed away 250 miles at northeast Honshu. The IDF send a delegation to build a small scale FFH in the format of clinic. Its clinic was located on the east coast in the town of Minami-Sanriku. It served mainly as a referral unit for diagnostic and medical treatment. It was staffed with 55 personnel. The structure of the FFH consisted of several wards: registration-triage and discharge, gynecology, internal medicine, laboratory, surgery, pediatrics, surgery, pharmacy, laboratory and imaging, and a logistics command center. Also, a team of 8 translators helped the FFH crew. In addition, there were an imaging crew equipped with ultrasound and X-ray, a hematology-microbiology-chemistry laboratory, and wireless services. Sources: [6, 15, 42]
Philippines	The typhoon Haiyan struck the Philippines on November 8, 2013. Five days after the event, an IDF team from Israel was assigned by the Philippines government to provide medical assistance to the city of Bogo, where a local hospital that serves more than 250,000 people was operating at partial capacity. The FFH team in the Philippines decided to combine its physical setup with the local structure and support the local medical staff with its experienced medical group, to provide maximum benefit and thereby create one integrated medical infrastructure. Although the IDF team had 25 physicians representing most medical subspecialties and first-class logistics support, they decided to relinquish sole decision-making authority and improvised to establish a model of cooperation with the local health care administrators. Sources: [31, 32, 37, 38]
Nepal	A 7.8 Richter magnitude earthquake struck Nepal on April 25, 2015. The IDF mission that established a field hospital in Kathmandu on April 29 consisted of 126 personnel including 45 physicians. They arrived with 100 tons of equipment and supplies, and capacity to treat 200 patients per day. It was largest IDF mission deployed overseas. Its wards included 2 operating rooms, 8-bed intensive care unit, trauma, obstetrics, gynecology, surgical, orthopedic, and imaging facility. Sources: [29, 30, 43]

Table 1.
Description of IDF disaster relief missions.

All missions were self-sufficient in terms of means of transportation, fuel, drinking water and food supplies, generators and electrical supply, communication systems, tents, kitchen and laundry accessories, and equipment for mechanical maintenance, as well as with means of physical security and preventive medicine. Medical equipment and supplies were based on standard gear of field hospitals stored in military warehouses, supplemented with specific items, medications and supplies tailored for specific mission characteristics [28]. All missions were equipped with standard units for field operations, with ventilators, monitors, and defibrillators, with oxygen supply, with X-ray and ultrasound machines and with a basic diagnostic laboratory (for blood counts, urinary chemistry analysis, microbiology cultures, blood smear staining, coagulation profile, blood gases analysis, serology, and with complementary facilities as needed, such as kits for HIV detection following accidental needle sticks by personnel. There was a limited supply and storage capacity for blood products and with the means for on-site collection, and

DOI: http://dx.doi.org/10.5772/intechopen.88214

Country	Armenia	Rwanda	Kosovo	Turkey (Adapazari)	Turkey (Duzce)	India	Haiti	Japan	Philippines	Nepal
Date (month, year)	Dec-88	Jul-94	Apr-99	Aug-99	Nov-99	Jan-01	Jan-10	Mar-11	Nov-13	Apr-15
Type of disaster	6.8 Richter earthquake	Rwandan refugees	Albanian refugees	7.6 Richter earthquake	7.2 Richter earthquake	7.7 Richter earthquake	7 Richter earthquake	9.0 Richter earthquake	Typhoon	7.8 Richter earthquake
Time until initiation of FFH	12 days		4 days	24–36 h	63 h	6 days	89 h	2 weeks	5 days	82 h
Duration of deployment	13 days	6 weeks	16 days	1 week	9 days	10 days	10 days	2 weeks	10 days	11 days
Number of casualties	25,000	Hundreds of thousands		2627	705	20,005	230,000	28,000	6300	9000
Number of injured	19,000	Hundreds of thousands		5084	3500	166,812	250,000	2800	28,000	23,000
Number of beds in FFH	25	50	35	80		30	72		80	60
Total number of patients	2400	6000	1560	1205	2230	1223	1111	400	2686	1668
Total personnel	34	110	76	102	100	97	100	55	147	126
Physicians	20	18	15	21	21		45	14	25	45
Nurses	3	21	7	10	13		27	7		29
Paramedics and medics	7	4	2	18	19		21			
Pharmacists	1	2	1	1			2	1	1	1
Radiology technicians	1	1	1	1	1		2	1	1	1
Laboratory technicians	1	1	1	1	1		3	2	1	1

Table 2.
Data on relief missions.

screening of blood products. Powered plasma, used for instance in Nepal, helped compensating for the limited storage capacity for blood products [29], and the use of ultrasound-guided nerve blocks for limb surgery saved turnover time and recovery from anesthesia [30].

An important result highlighted during the analysis was the ingredient of creativity needed in all missions with the variety of injuries and diseases they faced (disaster and non-disaster related). Crush injuries and traumatology in missions deployed to earthquake scenarios, epidemics in Rwanda, and later in Haiti, malnutrition and endemic diseases in both missions and in Kosovo, etc. In the mission to Philippines surgical interventions were considered in FFH for therapeutic, palliative, and diagnostic purposes of head and neck tumors [31]. Similarly, another example of improvisation during IDF mission to Philippines occurred when a child with a suspected brain abscess was successfully diagnosed and properly treated [32]. This complex heterogeneity required adaptations in equipment and supplies, not always foreseen, especially with altering clinical challenges. Other improvisations were the extended use of local or regional anesthesia over general anesthesia to shorten recovery periods, the primary abdominal closure with plastic infusion bags due to inflamed Shigella-related necrotizing enterocolitis requiring intestinal resection, blood donation by medical personnel to avoid HIV transmission to recipients in hyperendemic population in Rwanda and Haiti, or the use of protracted (days) ventilation with Ambu bag by hired personnel, in the case of continuous use of all available respirators. Another example of creativity is the self-production of orthopedic hardware, for instance the conversion of Steinman pins into Scentz screws with the aid of a local blacksmith and an engraving machine [33]. These screws underwent standard autoclave sterilization and proved effective in open fractures fixation in the Haiti mission.

5. Discussion

Although each and every disaster presents unique challenges for aid teams, numerous lessons can be derived from the ten IDF missions over the last three decades. After each mission, a rigorous post-mission comprehensive review was held in order to derive lessons about decisions regarding how to improve cooperation with local healthcare providers and foreign delegations.

In general, Israel policy has been to send a large delegation and allow both local and foreign medical personal to join its team [34, 35]. For example, in the IDF field hospital in Armenia, local medical staff that could not operate the destroyed local medical facilities was incorporated in the IDF FFH, enhancing efficacy, translating, utilizing available local facilities, such as sonography and laboratory equipment, and particularly by bridging cultural and professional gaps. Local logistic systems provided warmed housing, transportation and technical aid at the site of action. In Kosovo (1999), young Albanian students volunteered to provide translations, while in Haiti (2010) eight Columbian doctors and nurses joined the surgical teams, enabling around the clock surgeries at 3–4 operating tables.

At the IDF hospital in Rwanda [36], hired locals were used as translators, in preparing local food and in feeding patients, in the preparation of oral rehydration solutions, and in additional maintenance and logistic tasks. Locals were hired in Haiti, Nepal, and the Philippines as well. This help was especially important as the numbers of hospitalized patients increased over time. Incorporating a Dutch Medical Corps company that operated a rehabilitation/convalescence department for severely debilitated patients further expanded overall capacity to about 200 beds. In meetings held at the UNHCR headquarters, representatives of the various medical relief missions were briefed by CDC experts regarding epidemiology and susceptibility of prominent pathogens, exchanged clinical data and developed a working network of collaboration. Few examples are the conversion of the IDF FFH into a referral center for other medical facilities, the creation of an outflow tract for convalescing children without families in orphanages, a major contribution of a French Army Microbiology laboratory in the diagnosis and management of infectious diseases such as meningitis, and a help by various agencies in supplementing medical supplies and equipment at shortage.

Intact domestic third level medical facilities at the perimeter of the disaster settings enabled transfer of treated patients. This option occasionally offered stabilized critically wounded patients better critical care than in the field conditions. For instance, in Armenia (1988), a patient with ruptured viscera, shock and hypothermia was transported by a Russian Army helicopter, escorted by Israeli and local anesthesiologists to Yerevan, following a lifesaving urgent control of internal bleeding. Air transport of treated and stabilized patients to Macedonian hospitals in the Kosovo mission (1999), and to major hospitals in Ankara and Istanbul in the Turkish earthquake disasters (1999) helped maintaining the operating capacity of the FFH at the disaster settings. Similar cooperative pattern was adopted following the disaster in Nepal (2015), with the IDF FFH in Kathmandu working in collaboration with the Nepalese Birenda Army Hospital. In Bhuj, India (2001), an IDF Hercules airplane remained at hand, providing airlift of treated casualties to remote hospitals in India. In Haiti (2010), such patients were transferred to local primary care facilities to continue with postoperative care, facilitating coping with the never-ending stream of newly admitted patients. The best way to facilitate such cooperation among medical centers is through a centralized system such as the United Nations Disaster Assessment in this event, or the UNHCR headquarters in humanitarian aid to refugees.

A totally different type of collaboration might be the incorporation of the FFH medical staff within an overwhelmed and injured local medical facility. This approach was adopted in the FFH mission to the Philippines (2013) where it was decided to combine efforts with the local facility, creating one integrated medical infrastructure [37]. The IDF delegation was integrated with the Severo Verallo memorial district hospital, an urban healthcare facility with approximately 80 beds, which was understaffed and had limited resources. The IDF 25 physicians representing most medical subspecialties, with the additional medical personnel worked under the medical and administrative direction of the local health care directors, while the logistic staff assisted the repair of the local hospital, restoring electricity, and providing much-needed supplies and equipment such as a mobile X-ray machine and an autoclave. Open discussions, held to establish clear lines of responsibility and co-sharing of tasks, helped in building trust and cooperation [38].

The mission's termination timing depended on the resolution of the disastrous event, for instance the termination of influx of patients removed from ruins in earthquake Turkish disasters, or the control of diarrheal epidemics among Rwandan refugees by the installation of appropriate water and food supplies and sanitation. Another important factor is the restoration of local health systems, as occurred in Armenia or the Philippines, or the establishment of appropriate long-lasting substitute services such as a Norwegian field hospital that settled at Goma, Zair, or the arrival of the USNS Comfort floating hospital, and other medical facilities operated by the Red cross and the University of Miami in the Haitian disaster. In such settings a handing over procedure of hospitalized patients was carried out, with their available medical data. Some convalescing but fully incapacitated patients were handed over to other non-medical local humanitarian facilities such as monasteries and orphanages. The termination of mission was coordinated with and orchestrated by the local health authorities in order to ensure continuity of operations by local medical staff or newly arrived substitutes. In most cases, supplies and equipment were handed over to local health systems under the direction of local authorities.

While the selection and incorporation of drafted highly qualified medical personnel within the military framework, characteristic for the Israeli FFH model, provides excellent medical performance customized for the specific mission, this restricts the longevity of the mission, as drafted personnel are expected to resume their civil work within a relatively short period of time. Thus, most missions lasted 2–3 weeks, only. Nevertheless, as happened in the Rwandan disaster, and in other missions addressing disasters in Cambodia and in Ethiopia, substituting teams of medical experts and additional personnel were created with changing operative shifts at 2–4 weeks intervals [39]. This enabled protracted mission activity, as required.

6. Conclusions

In conclusion, our study provides comprehensive review of ten missions conducted by the IDF over the last three decades. **Table 3** summarizes insights emerging from our research for future relief missions. The uniqueness of our study is that we investigated the response to different types of disasters, with some of the humanitarian missions sent in response to natural disasters (earthquake, tsunami, or typhoon) while others were delivered to war zones and ethnic clash terrains.

1.	An advanced team is crucial for defining needs, expectations, priorities, and identifying risks, as well as facilitating legal details with local authorities
2.	Swift deployment providing adaptive operative flexibility is maintained by delegation's multi-disciplinary heterogeneity of personnel, and readiness for improvisations
3.	Coordination with both the local health system and other aid organizations operations in disaster area is essential
4.	It is imperative to be aware and respect the national culture differences between an aid mission and the affected country
5.	Field hospital must be entirely self-sufficient (transportation, energy, food, water, equipment and supplies)
6.	Providing security to field hospital may be necessary in conflict areas
7.	The contribution of translators and local health employees is significant
8.	Integration of volunteer teams from other countries into field hospital can fill lack of human resource and improve operations
9.	The optimal operative period is 2–3 weeks. Substitutions and supplementary airborne logistics are required for longer missions
10.	Standardization of procedures is essential in order to optimize medical response
11.	After few days, most of medical activity becomes non-urgent treatment of population
12.	Communication devices, Information systems, and electronic medical data storage and handling records improve efficiency of field hospital
13.	Before departure back to home country, the delegation should coordinate with local authorities the transfer of authority over the FFH facility, equipment, and supplies in order to ensure continuity of operations by local medical staff
14.	Ethical issues pertaining to treatment of patients and their families in disaster area must be taken into consideration before mission deployment

Table 3.
Insights emerging from our study for future relief missions.

More specifically, the cases we describe in Rwanda and Kosovo contribute to the literature because they demonstrate the usefulness of FFHs not only for natural disasters but also in situations of civil war. The ten missions varied also in their geographical distance from the home base in Israel, which impacted arrival time. While some of the disasters where located in developing countries such as Haiti, others occurred in highly developed countries such as Japan. The objective of the missions determined the duration of stay. For example, in the earthquake disasters in Turkey, FFHs were designed to assist in the acute care of casualties for a few days, only, within the period of time of recovering survivors trapped under the ruble of collapsed buildings, while at Rwanda, the goal was to participate in protracted, large scale lethal epidemics among displaced population fleeing a civil war.

A crucial avenue of research concerns the ethical issues humanitarian operations face, such as which patients to admit, given the limited capacity of hospital beds and other resources [40, 41]. For instance, it was noted based on experience in Japan and Haiti that premature deliveries, low-birth-weight neonates, and other complications with increased risk of infection and blood-loss increase at disaster areas. Because survival rates of low-birth weight neonates delivered in a disaster environment are diminished, the dilemma of whether to impose a minimum weight threshold for preterm neonates to receive treatment is an ethical issue, which obstetrics and gynecology teams should be prepared to deal with [42]. Another example, during Nepal mission it was found that procedural sedation and analgesia (PSA) should be a priority when treating pediatric victims of disaster since they are prone to psychological distress secondary to the traumatic event [43].

In sum, this study has highlighted the importance of studying the fruitful collaboration between military and civilian organizations in the establishment of field hospitals. We hope that the future research avenues we sketched will motivate other scholars to engage in this field to prepare the global community to deal with major disaster relief operations.

Author details

Michael Naor
School of Business Administration, The Hebrew University of Jerusalem, Jerusalem, Israel

*Address all correspondence to: michael.naor@mail.huji.ac.il

References

[1] Peleg K, Reuveni H, Stein M. Earthquake disasters: Lessons to be learned. Israel Medical Association Journal. 2002;**4**:361-365

[2] Rokach A, Nemet D, Dudkiewicz M, Albalansi A, Pinkert M, Schwartz D, et al. Advanced rescue techniques: Lessons learned from the collapse of a building in Nairobi, Kenya. Disasters. 2009;**33**(2):171-179

[3] WHO/PAHO. Guidelines for the Use of Foreign Field Hospitals in the Aftermath of Sudden-Impact Disaster. Department of Emergency and Humanitarian Action, the World Health Organization; Area of Emergency Preparedness and Disaster Relief; 2003

[4] Ran Y, Hadad E, Daher S, Ganor O, Yegorov Y, Katzenell U, et al. Triage and air evacuation strategy for mass casualty events: A model based on combat experience. Military Medicine. 2011;**176**(6):647-651

[5] Kreiss Y, Merin O, Peleg K, Levy G, Sagi R, Abargel A, et al. Early disaster response in Haiti: The Israeli field hospital experience. Annals of Internal Medicine. 2010;**153**(1):45-48

[6] Finestone AS, Levy G, Bar-Dayan Y. Telecommunications in Israeli field hospital deployed to three crisis zones. Disasters. 2014;**38**(4):833-845

[7] Heyman S, Eldad A, Wiener M. Airborne field hospital in disaster area: Lessons from Armenia (1988) and Rwanda (1994). Prehospital and Disaster Medicine. 1998a;**13**(1):14-21

[8] Bar-On E, Abargel A, Peleg K, Kreiss Y. Coping with the challenges of early disaster response: 24 years of field hospital experience after earthquake. Disaster Medicine and Public Health Preparedness. 2013;7(5):491-498

[9] Farfel A, Assa A, Amir I, Bader T, Bartal C, Kreiss Y, et al. Haiti earthquake 2010: A field hospital pediatric perspective. European Journal of Pediatrics. 2011;**170**:519-525

[10] Levy G, Blumberg N, Kreiss Y, Ash N, Merin O. Application of information technology within a field hospital deployment following the January 2010 Haiti earthquake disaster. Journal of American Medical Association. 2010;**17**:626-630

[11] Rosen Y, Gurman P, Verna E, Elman N, Laor E. Rapidly-deployed small tent hospitals: Lessons from the earthquake in Haiti. Prehospital & Disaster Medicine. 2012;**27**(3):1-2

[12] Adini B, Peleg K, Cohen R, Laor D. A national system for disseminating information on victims during mass casualty incidents. Disasters. 2010;**34**(2):542-551

[13] Naor M, Linderman K, Schroeder RG. The globalization of operations in eastern and Western countries: Unpacking the relationship between national and organizational culture and its impact on manufacturing performance. Journal of Operations Management. 2010;**28**(3):194-205

[14] Hartal M, Yavnai N, Yaniv G, Gertz D, Fleshler E, Kreiss Y. Old challenges and new perspectives on developing military physicians: The first 4 years of the new Israeli model. Military Medicine. 2016;**181**(2):129-135

[15] Merin O, Blumberg N, Raveh D, Bar A, Nishizawa M, Cohen-Marom O. Global responsibility in mass casualty events: The Israeli experience in Japan. American Journal of Disaster Medicine. 2012;**7**(1):61-64

[16] Margalit G, Rosen Y, Tekes-Manova D, Golan M, Benedek P, Levy Y, et al.

Recommendations for nursing requirements at a field hospital, based on Israel defense forces field hospital at the earthquake disaster in Turkey—August 1999. Accident and Emergency Nursing. 2002;**10**(4):217-220

[17] Maayan S, Marks N, Viterbro A, Zeide Y, Morag A, Neil L, et al. HIV infection and susceptibility to epidemic bacterial infections among Rwandan refugees. International Journal of Infectious Diseases. 1997;**1**(4):199-201

[18] Heyman SN, Ginosar Y, Niel L, Amir J, Marx N, Shapiro M, et al. Meningococcal meningitis among Rwandan refugees: Diagnosis, management, and outcome in a field hospital. International Journal of Infectious Diseases. 1998b;**2**(3):137-142

[19] Halpern P, Rosen B, Carasso S, Sorkine P, Wolf Y, Benedek P, et al. Intensive care in a field hospital in an urban disaster area: Lessons from the august 1999 earthquake in Turkey. Critical Care Medicine. 2003;**31**(5):1410-1414

[20] Wolf Y, Bar Dayan Y, Mankuta D, Finestone A, Onn E, Morgenstern D, et al. An earthquake disaster in Turkey: Assessment of the need for plastic surgery services in a crisis intervention field hospital. Plastic & Reconstructive Surgery. 2001;**107**(1):163-170

[21] Merin O, Miskin IN, Lin G, Wiser I, Kreiss Y. Triage in mass-casualty events: The Haitian experience. Prehospital & Disaster Medicine. 2011;**26**(5):386-390

[22] Yitzhak A, Sagi R, Bader T, Assa A, Farfel A, Merin O, et al. Pediatric ventilation in a disaster: Clinical and ethical decision making. Critical Care Medicine. 2012;**40**(2):603-607

[23] Bar-Dayan Y, Beard P, Mankuta D, Finestone A, Wolf Y, Gruzman C, et al. An earthquake disaster in Turkey: An overview of the experience of the Israeli defense forces field hospital in Adapazari. Disasters. 2000;**24**(3):262-270

[24] Bar-Dayan Y, Leiba A, Beard P, Mankuta D, Engelhart D, Beer Y, et al. A multidisciplinary field hospital as a substitute for medical hospital care in the aftermath of an earthquake: The experience of the Israeli defense forces field hospital in Duzce, Turkey, 1999. Prehospital and Disaster Medicine. 2005;**20**(2):103-106

[25] Heyman SN, Ginosar Y, Shapiro M, Kluger Y, Marx N, Mayyan S. Diarrheal epidemics among Rwandan refugees in 1994: Management and outcome in a field hospital. Journal of Clinical Gastroenterology. 1997;**25**(4):595-601

[26] Amital H, Alkan JA, Kriess I, Levi Y. Israeli defense forces medical corps humanitarian mission for Kosovo's refugees. Prehospital and Disaster Medicine. 2003;**18**(4):301-305

[27] Adler J, Bodner E, Stein SB, Goldfarb J, Engelhard D, Naparstek J, et al. Medical Mission to a refugee camp in Thailand. Disasters. 1981;**5**(1):23-31

[28] Finestone AS, Bar-Dayan Y, Wolf Y, Stein M, Tearosh J, Zaide Y, et al. Diagnostic medical auxiliary equipment in a field hospital: Experience from the Israeli delegation to the site of the Turkish earthquake at Adapazari. Military Medicine. 2001;**166**(7):637-640

[29] Merin O, Yitzhak A, Bader T. Medicine in disaster area: Lessons from 2015 earthquake in Nepal. JAMA Internal Medicine. 2015:E1-E2

[30] Lehavi A, Meroz Y, Maryanovsky M, Merin O, Blumberg N, Bar-On E, et al. Role of regional anaesthesia in disaster medicine: Field hospital experience after the 2015 Nepal earthquake. European Journal of Anaesthesiology. 2016;**33**(5):312-313

[31] Marom T, Merin O, Segal D, Tsumi E, Lin G, Erlich T. Ethical and clinical dilemmas in patients with head and neck tumors visiting a field hospital in the Philippines. American Journal of Disaster Medicine. 2014;**9**(3):1-8

[32] Weiser G, Mendlovic J, Dagan D, Albukrek D, Shpriz M, Merin O. Case report: A brain abscess in a disaster zone—Beyond the call of duty. Disaster and Military Medicine. 2015;**1**(13):1-3

[33] Lin G, Lavon H, Gelfond R, Abargel A, Merin O. Hard times call for creative solutions: Medical improvisations at the Israel field hospital in Haiti. American Journal of Disaster Medicine. 2010;**5**(3):188-192

[34] Bar-Dayan Y. International collaboration in disaster medicine – It is time for the "big step" in disaster preparedness. Prehospital and Disaster Medicine. 2008;**23**(4):280-281

[35] Peleg K, Kreiss Y, Ash N, Lipsky AM. Optimizing medical response to large-scale disasters: The ad hoc collaborative health care system. Annals of Surgery. 2010;**253**(2):421-423

[36] Heyman SN, Horovitz J, Nehama H, Sofer S, Orbach J, Amir Y, et al. Sudden death during fluid resuscitation: A lesson from Rwanda. Lancet. 1994;**344**:1509-1510

[37] Albukrek D, Mendlovic J, Marom T. Typhoon Haiyan disaster in the Philippines: Pediatric field hospital perspectives. Emergency Medicine Journal. 2014;**31**(12):951-953

[38] Merin O, Kreiss Y, Lin G, Pras E, Dagan D. Collaboration in response to disaster—Typhoon Yolanda and an integrative model. New England Journal of Medicine. 2014;**370**:1183-1184

[39] Kassis I, Lak L, Adler J, Rinat C, Shazberg G, Fekede Tewade D, et al. Medical relief operation to rural northern Ethiopia: Addressing an ongoing disaster. Israel Medical Association Journal. 2001;**3**:772-777

[40] Merin O, Ash N, Levy G, Schwaber MJ, Kreiss Y. The Israeli field hospital in Haiti: Ethical dilemmas in early disaster response. New England Journal of Medicine. 2010;**362**(1):e38

[41] Glassberg E, Bader T, Nadler R, Benov A, Zarka S, Kreiss Y. When humanitarian trumps politics. Israel Medical Association Journal. 2015;**17**:339-340

[42] Pinkert M, Dar S, Goldberg D, Abargel A, Cohen-Marom O, Kreiss Y. Lessons learned from an obstetrics and gynecology field hospital response to natural disasters. Obstetrics and Gynecology. 2013;**122**(3):532-536

[43] Weiser G, Ilan U, Mendlovic J, Bader T, Shavit I. Procedural sedation and analgesia in the emergency room of a field hospital after the Nepal earthquake. Emergency Medicine Journal. 2016;**33**(10):745-747

Chapter 4

A New Real-Time Flight Simulator for Military Training Using Mechatronics and Cyber-Physical System Methods

César Villacís, Walter Fuertes, Luis Escobar, Fabián Romero and Santiago Chamorro

Abstract

So far, the aeronautical industry has developed flight simulators and space disorientation with high costs. This chapter focuses on the design and implementation process of a low-cost real-time flight simulator for the training of armed force pilots using mathematical models of flight physics. To address such concern, the mathematical models of a Cessna type aircraft have been developed. This has been followed by a flight simulator, which operated with a new construction using a Stewart scale platform and operated by a joystick. Specifically, the simulator has been developed using an approximation of a physical cyber-system and a mechatronic design methodology that consists of mechanical, electrical and electronic elements that control the Stewart platform with three degrees of freedom. Based on software engineering, the algorithms of mathematical and physical models have been developed. These have been used to create an interactive flight simulator of an aircraft based on the Unity 3D game engine platform. The performance of the algorithms has been evaluated, using threads and processes to handle the communication and data transmission of the flight simulator to the Stewart platform. The evaluation of the developed simulator has been validated with professional pilots drilled with the Microsoft Flight Simulator. The results demonstrated that this flight simulator stimulates the development of skills and abilities for the maneuver and control of an aircraft.

Keywords: flight simulator, mathematical models of aircraft, flight training, Stewart platform, real time simulation, mechatronics, cyber-physical system, kinematic control

1. Introduction

Modern flight simulators meet two main aviation objectives: (1) to provide pilot training at the instructor's level, and at the student's level to learn to fly and to earn virtual flight hours that are useful for flying real aircrafts and (2) to simulate normal flight conditions, as well as adverse situations and spatial disorientation such as navigation instrument faults, power losses, loss of control of the aircraft, confusion illusion of references, illusion of the effect of black holes, among others, that would be dangerous and even catastrophic in a real flight; so, they must be well analyzed, controlled, and learned.

Real flight simulators with full movement generate movements and images where pilots feel an almost 100% level of realism of what would happen in a real plane. These simulators combine a series of technological aspects such as the Stewart platforms that reproduce real-time movements of the simulator software at hardware level and allow stimulating the visual and vestibular system of the pilots, reaching a maximum level of knowledge of various types of favorable and adverse situations and spatial illusions.

The main objective of this research was to design and build a flight simulator as a cyber-physical system, both at the software level and at the hardware level; based on mathematical models and programming algorithms that allow us to recreate a Cessna 172 type airplane in a virtual world and to prove its correct functioning. For that purpose, the Unity 3D gaming engine has been used as a developmental tool, together with the LabVIEW graphical programming environment to access the hardware and data information of the built-in Stewart platform with three degrees of freedom.

The flight simulator as a cyber-physical system composed of software and hardware was evaluated in terms of its functionality with a group of Ecuadorian aviation pilots who met basic training flight hours with the Cessna 172 and the ENAER T-35 Pillan, and also met virtual flight hours using a personalized license from Microsoft Flight Simulator.

The main contributions of this study were: (a) design of a mathematical model of front velocity, vertical velocity and lateral velocity of an airplane, as well as the application of the physical forces involved in an airplane such as lift, weight, thrust, and drag; (b) implementation of a dynamic Cyber-physical system, where the Mechatronic system was part of it and consisted of a software flight simulator programmed with the UNITY 3D framework. It allowed to include 3D objects such as terrains, buildings, airplanes, cities, sea, etc.; as well as, to model the cockpit in 3D as it looks in the reality and a Stewart platform (hardware) to scale with three degrees of freedom that reproduces the movements of the simulator plane in relation to the roll, pitch, and yaw, operated by a joystick.

This research has been organized as follows: Section 2 talks about the mechatronics systems design based in V-Model. Then, Section 3 explains about mathematical foundations of Stewart-Gough Platform with 3-DOF. Later, the flight simulator construction has been clarified in Section 4. Next, Section 5 presents the experimental results. Finally, Section 6 gives the conclusions and future work.

2. Mechatronics systems design based in V-Model

2.1 Philosophy

The philosophy of designing mechatronic systems has evolved over the years along with its definition, applications, and boundaries. Since its origins, mechatronics has tried to analyze systems holistically, with a synergic point of view integrating heterogeneous components (mechanics, electronics, and informatics) and subsystems to create more complex systems. This method requires a development that delivers products independently of their domains.

The V-Model allows the design and development of complex mechatronic systems with an interdisciplinary approach, where the VDI guideline 2206 can be applied to obtain systems that are more flexible and adaptable to the needs of users [1]. This model considers different factors, components, and the synergistic behavior of the different parts of a mechatronic product and its integration.

The implementation of mechatronic systems has a great reception in the military area. The development of vehicles is one of the most obvious examples due to the high variety of functions that must be fulfilled, according to [2] the idea of using mechatronic systems in the design of combat vehicles arrived with the twenty-first century when the basic architecture of electric and electronic components was developed, a clear example of these systems can be observed in [3, 4].

The framework given for that methodology has as main objective to generate a product based on the client requirements. In order to accomplish their requirements, the characteristic method to design and develop mechatronic systems is composed of four phases: (1) requirements phase—requirements and specifications analysis is customer oriented and defines the start and end of the project; (2) functional characteristics phase—the functions that are directly or indirectly visible by the user are defined; (3) design phase—the hardware and software components are defined, as well as the architecture of the system; (4) implementation phase—all the unit elements or system programming modules are developed [5]. Then, in the present study, this methodology contributed in the design and implementation of hardware for a simulator.

2.2 Cyber-physical system (CPS)

Cyber-physical system (CPS) is a new concept around the revolution of Information and Communications Technology (ICT) and Embedded Systems, and it refers to the integration of computing, data networks and physical processes to become intelligent objects that can cooperate with each other, forming distributed and autonomous systems. A CPS can include several disciplines related to software (Systems Engineering, Computation, Communication, and Control) and hardware (Industrial Engineering, Mechanical, Electrical, and Electronic) [4].

Nowadays, thanks to the internet of things, there is a greater development of products that have computing and communication. These products need an intimate coupling between the cyber and physical components and will be presented in the nano at large-scale world. CPS may be considered a confluence of embedded, real-time and distributed sensor systems as well as control [6, 7].

As it can be seen in the military [8] the modeling of present physical systems has several challenges, such is the case of fuel tanks in airplanes. Maintaining a correct monitoring of the level of fuel is a complex job since it requires the use of several sensors and the consideration of several external factors, but when generating a solution with CPS has advantages in the monitoring of it that helps to prevent accidents such as the case of Air Transat Flight on August 23, 2001; where due to a maintenance failure, the aircraft was left without fuel for the landing. This type of systems can be implemented in various systems of military vehicles such as jet aircraft, helicopters, etc.

By having a CPS, it is possible not only to monitor, but to implement an autonomous control; this can be done in any sector such as factories, transportation, aerospace, buildings and environmental control, process control, critical infrastructure, and healthcare. As indicated by [9], CPS projections will have a great impact in a wide variety of areas, and [10] states that Networked autonomous vehicles could dramatically enhance the effectiveness of the military and could offer substantially more effective disaster recovery techniques.

The biggest opportunity for implementing CPS is given by the opportunity to decrease costs and simultaneously increase capacity of sensors and actuators; in addition to the access to high capacity, smaller formed computing devices, wireless communication, increased bandwidth and continuous improvements in energy. The advantages and opportunities offered by the cyber-physical systems (CPS) have allowed innovation and improvement of the principles of engineering as Rajkumar mentioned in his research [9]. One of the main challenges that arises in the design and development of cyber-physical systems is the development of new methods of science and system design engineering to obtain a CPS that is compatible, reliable, integrated and synergistic in all the five-layers of functionality architecture with other cyber-physical systems. This research contributes to the design, development and implementation of a CPS following engineering principles.

2.3 Mechatronics and CPS

Naturally CPS and mechatronics systems are different, each one follows different goals, but its subtle differences characteristics make those systems complementary; since CPS consider the mechatronic system as an integral part of them. **Table 1** shows the differences between mechatronic systems and CPS.

According to the works of [11, 12], the cyber-physical systems are composed of two parts.

2.3.1 Software (Cyber)

This component has as areas of influence the Systems Engineering, Computation Engineering, Communication Engineering, and Control Engineering; all of these supported by the Application Lifecycle Management (ALM).

2.3.2 Hardware (Physical)

This component refers to the mechatronic itself, which has as areas of influence Industrial Engineering, Mechanical Engineering, Electrical Engineering, and Electronic Engineering; all of these supported by the Product Lifecycle Management (PLM).

Parameters	Mechatronics	CPS
Maintainability/availability	+	
Scalability		++
Stability		+
Robustness		++
Efficiency	+	+
Autonomous		++
Energy efficiency	+	++
Safety	++	++
Compactness	+	
Reliability	++	+
Accuracy	++	+
Communication	++	++

Table 1.
Differences between mechatronic systems and CPS.

In this research, it has been designed as a complex system applying the approach of cyber-physical systems, where all the hardware components designed complied with Mechatronic processes and the software components designed complied with software engineering processes. Traditionally, practitioners of the mechatronic design philosophy had focused in the creation of a specialized architecture depending on the proposed solutions, while CPS are per se more distributed systems. Subsequently, in this study we provide the integration of hardware simulator with software application in order to obtain a complex dynamic system.

3. Stewart-Gough platform with 3-DOF, architecture and mathematical foundation

Based on the fact that for the construction of real flight simulators with full movement contemplated as cyber-physical systems which include hardware and software, the classic platforms of Stewart Gough of 6-DOF have been used. In the current project, a platform of Stewart Gough has been designed modified to be conducted with a 3-DOF in a limited and low cost manner that restricts the superior ability to perform linear movements on the β (x, y, z) plane. Nonetheless, it will maintain its ability to lead any controlled orientation during an airplane flight in order to receive various effects of spatial disorientation to which the pilots are exposed in real life. Among them are the following that have been simulated in this project: (a) illusion of track; (b) approximation illusions; and (c) illusions due to land degradation or fusion. Through these robotic hardware and software platforms, both military and private pilots may be trained allowing them to develop spatial orientation skills in order to avoid potential catastrophic accidents.

In this research, the architecture proposed by Hunt [13] has been applied, which is known as the 3-RPS parallel robot. Such architecture consists of three identical RPS legs, whose lengths are changed with prismatic joints, while the platform moves with 3-DOF, as illustrated in **Figure 1**.

The basic equations used of the Stewart platform with 3-DOF according to [14–18], have been the following:

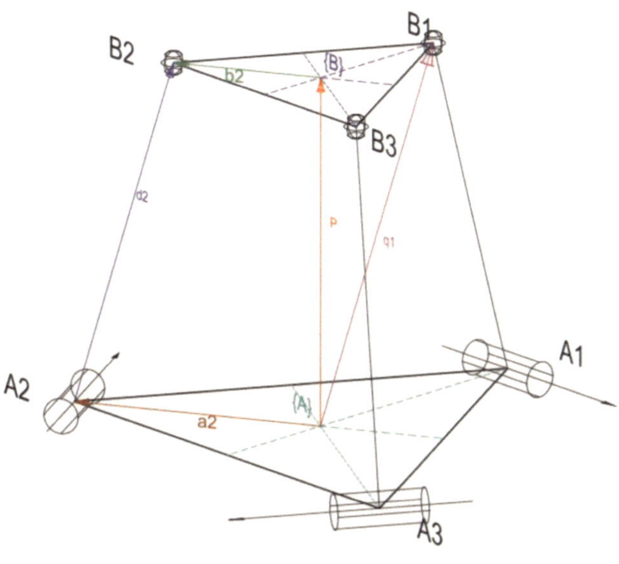

Figure 1.
Schematic illustration of the design of the platform.

Equations of vectors a_1, a_2 and a_3:

$$
\begin{aligned}
a_1 &= [g \ \ 0 \ \ 0]^T \\
a_2 &= \left[-\tfrac{1}{2}g \ \ \tfrac{\sqrt{3}}{2}g \ \ 0\right]^T \\
a_3 &= \left[-\tfrac{1}{2}g \ \ -\tfrac{\sqrt{3}}{2}g \ \ 0\right]^T
\end{aligned} \tag{1}
$$

Equations of the vectors b_1, b_2 and b_3:

$$
\begin{aligned}
b_1 &= [h \ \ 0 \ \ 0]^T \\
b_2 &= \left[-\tfrac{1}{2}h \ \ \tfrac{\sqrt{3}}{2}h \ \ 0\right]^T \\
b_3 &= \left[-\tfrac{1}{2}h \ \ -\tfrac{\sqrt{3}}{2}h \ \ 0\right]^T
\end{aligned} \tag{2}
$$

Equations of the prismatic joint represented by three points, namely q_1, q_2 and q_3:

$$q_i = p + {}^{A}R_B \bullet b_i, \ i = 1, 2, 3 \tag{3}$$

Equation of the position of vector p:

$$p = \left[p_x \ \ p_y \ \ p_z\right]^T \tag{4}$$

Equation of the rotation matrix ${}^{A}R_B$.

$$ {}^{A}R_B = \begin{pmatrix} u_x & v_x & w_x \\ u_y & v_y & w_y \\ u_z & v_z & w_z \end{pmatrix} \tag{5}$$

The equation of the joint q_1 whose procedure is equivalent for q_2 and q_3:

$$q_1 = \begin{bmatrix} p_x \\ p_y \\ p_z \end{bmatrix} + \begin{bmatrix} u_x & v_x & w_x \\ u_y & v_y & w_y \\ u_z & v_z & w_z \end{bmatrix} \bullet \begin{bmatrix} h \\ 0 \\ 0 \end{bmatrix} = \begin{bmatrix} p_x + h \bullet u_x \\ p_y + h \bullet u_y \\ p_z + h \bullet u_z \end{bmatrix} \tag{6}$$

4. Flight simulator construction

4.1 Architecture diagram of the flight simulator system

Figure 2 shows the architecture of the system consists of two layers.

4.1.1 First layer

This layer consists of three subsystems that are: (1) subsystem of the flight simulator control, which handles the logic of the virtual flight of the plane applying mathematical, Physics, and aeronautical models; (2) communication subsystem, which manages the connection with the second layer using sockets, the TCP/IP protocol, and flat files containing the information of the roll, pitch and yaw, corresponding to the movement of the plane; and (3) graphical user interface

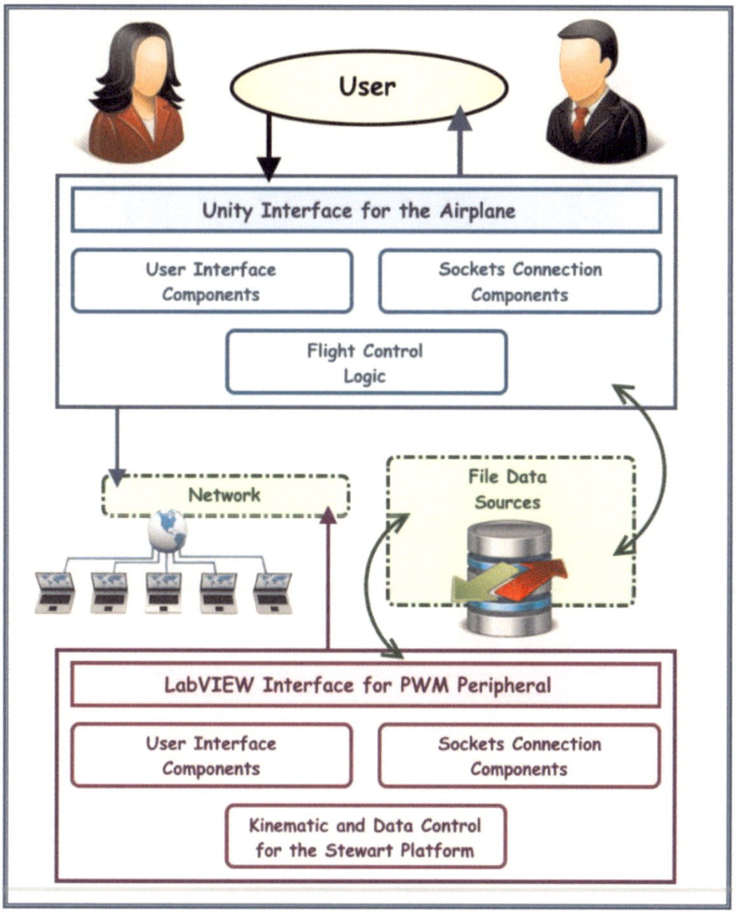

Figure 2.
Architecture diagram of the flight simulator system.

subsystem, which manages the components and assets of Unity 3D to create the virtual world that corresponds to the city of Manta, which is the destination of the plane.

4.1.2 Second layer

This layer consists of three subsystems that are: (1) communication subsystem—manages the connection with the first layer using sockets and reads the flat file with the information of the roll, pitch and yaw of the plane; (2) kinematic control subsystem—obtains the information corresponding to the roll, pitch and yaw of the airplane to be able to reproduce its movements in the Stewart platform with three degrees of freedom using servo motors; and (3) graphical user interface subsystem—manages the LabView components to view in real time the roll, pitch, and yaw values of the aircraft to operate the Stewart platform.

4.2 Flight simulator dynamic system

In this research, we propose a model of the dynamic flight system, which is based on the models proposed by [6, 19, 20]. **Figure 3** shows the interaction between the pilot, the airplane and the Stewart platform with three degrees of freedom. The pilot operates the flight simulator with a joystick, where it controls

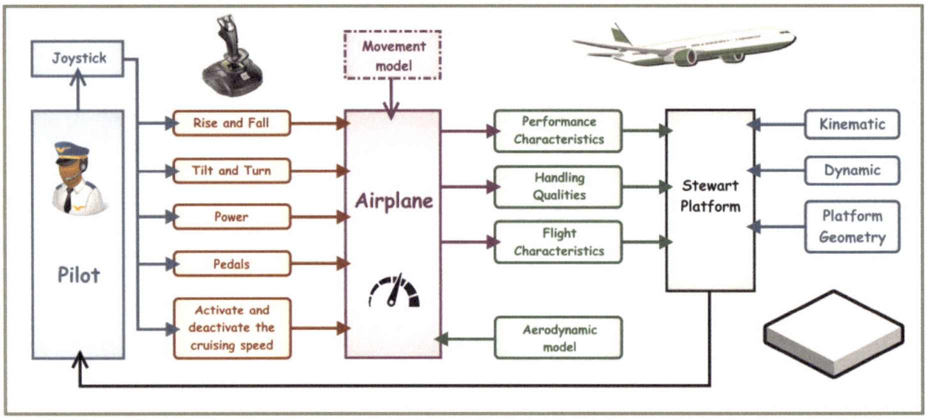

Figure 3.
Proposed flight simulator dynamic system.

the following components: (1) rise and fall of the airplane (range: [−45, 45]); (2) tilt and turn of the airplane (range: [−30 to 30°]); (3) power of the airplane (range: [−1, 1]); (4) pedals of the airplane; and (5) activation and deactivation of the airplane's cruising speed. The Stewart platform reproduces the roll, pitch and yaw movements generated by the flight simulator which is programmed with mathematical models of the flight physics.

For the operation of the Stewart platform with three degrees of freedom, we used a Multifunction DAQ Tested to Data Acquisition Card with pulse-width modulation (PWM) of Texas Instruments brand, which was programmed with LabVIEW to operate the kinematics and dynamics of the system considering the triangular geometry of the platform. This card manages: (1) the performance characteristics, such as control of the power and acceleration of the airplane; (2) handling qualities, such as the control of the speed of advance and rotation of the airplane; and (3) flight characteristics, such as airplane short take-off and landing functions, altitude limit alarm, ground proximity alarm, and landing gear control alarm.

4.3 Design and development of the mechatronic system

The design and construction of the mechatronic product as a complex system was carried out based on Model V, as shown in **Figure 4**. The following tasks have been accomplished according to different areas of knowledge: (1) Mechanical Engineering—This area was applied to adapt and build the mechanical elements of the Stewart platform with three translational degrees of freedom, considering that the proposed dynamic model has no friction and that the kinematic chains are symmetrical and thin; (2) Electrical and Electronic Engineering—This area was applied to control the Stewart platform using servo motors to simplify the inertia and flexibility of the kinematic chains; and (3) Information Technologies—This area was applied to program the data acquisition card with pulse-width modulation (PWM) using LabVIEW and to be able to operate the kinematics and the dynamics of the system with triangular geometry, which reproduces the roll, pitch and yaw movements of the flight simulator.

The code of the control program was based on the block diagrams generated by LabVIEW, for which the data referring to the roll, pitch, and yaw that the flight simulator outputs is read from a flat file. This data is interpreted numerically and processed to be sent to the servo motors that operate the Stewart platform and allow

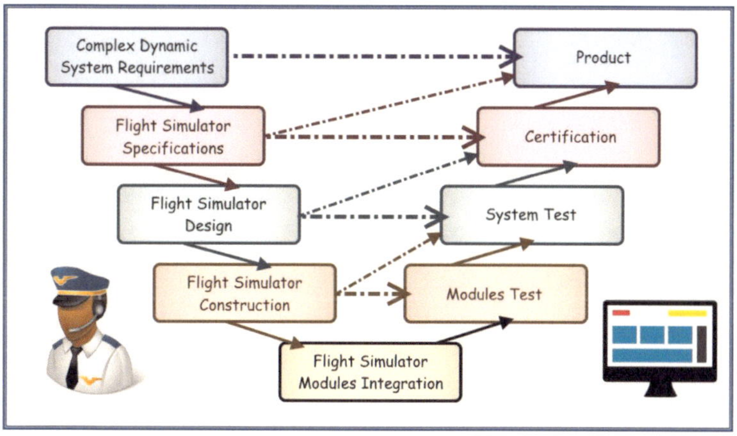

Figure 4.
Application of the V-Model for the flight simulator development.

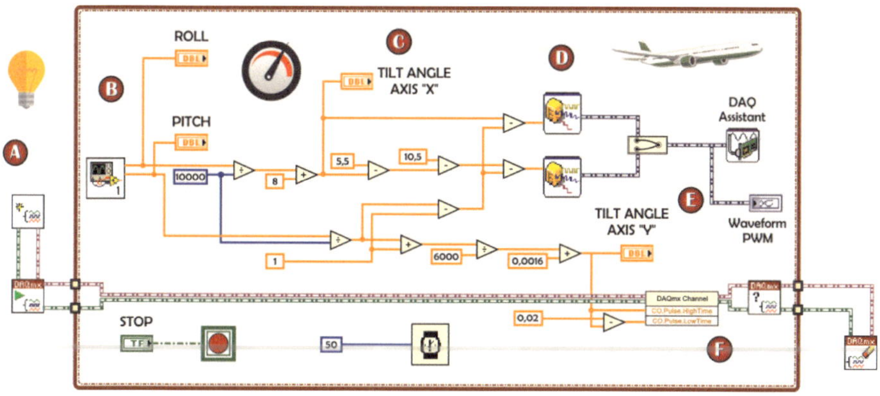

Figure 5.
Block diagram source code for controlling the Stewart platform in LabView.

the movements of the flight simulator to be reproduced. In **Figure 5**, you can see the data flow as described below: (1) programming a pulse-width modulation (PWM) channel; (2) programming functions of downloading and uploading roll, pitch and yaw data that come from the flight simulator; (3) programming conversion of roll, pitch and yaw values into pulse width modulation (PWM) signals to be able to move the servo motors; (4) programming the generation of the square signal by assigning a constant value to the period corresponding to 50 Hz; (5) programming the connection between the processed data with the data acquisition card with pulse-width modulation (PWM) signals to control the servo motors 1 and 3; (6) programming the connection between the processed data with the data acquisition card with pulse-width modulation (PWM) signals to control the servo motor 2.

Figure 6 shows the flow of information from computer 1 (where the flight simulator is installed) to computer 2 (where LabVIEW is installed). The flight simulator saves the data referring to the roll, pitch and yaw in a text file. Then computer 2 accesses this text file through a point-to-point connection using the TCP/IP protocol. Computer 2 receives the data from the text file and operates the control software that works with a data acquisition card with pulse-width modulation (PWM) signals, connected through a USB port to the computer and drives the servo motors of the Stewart platform with three degrees of freedom. Finally, there is a 5 V power supply that powers the servo motors. The mathematical model of the

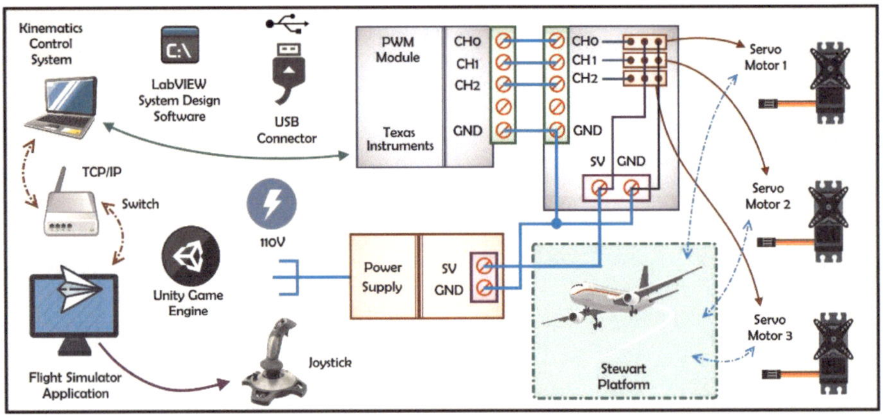

Figure 6.
Infographic of the electric and mechanical diagram of Stewart platform.

flight simulator was programmed with the UNITY 3D framework in computer 1, where the frontal velocity, vertical velocity and lateral velocity are considered allowing the airplane to move in the virtual world as it would be in the real world.

4.4 Mathematical model

In this research, we developed three basic mathematical models for the correct operation of the flight simulator, considering that an aircraft is a mass subjected to different forces such as weight (W), lift (L), thrust (T), and drag (D). In this sense, an analysis of each of these forces and the causes that produce them was made. This allowed the flight simulator to work according to the physical laws that are produced in the air and that the pilot know to properly control of an airplane.

4.4.1 Mathematical model of the front velocity

4.4.1.1 Front velocity

According to [21–23], the thrust force generates the frontal displacement of the airplane. This displacement can be explained using the classical physical laws considering to F_t like a punctual force applied to the mass of the airplane. The following equation is obtained:

$$V_x = V_{ix} + \left(\frac{V_e \cdot k_h}{m_a}\right) \cdot t \tag{7}$$

where F_t is the thrust force; V_e is the engine speed or helix; k_h is the thrust constant; m_a is the mass of the airplane; V_x is the front speed; V_{ix} is the initial velocity of the airplane; and t is the time.

4.4.2 Mathematical model of the vertical velocity

According to [21–23], the lift force generates the vertical displacement of the airplane. This displacement could be explained by using the classical physical laws considering the lift force like a punctual force applied to the mass of the airplane. The following equation is obtained:

$$V_V = V_{vi} \frac{t \cdot \left(\left(\rho \left(V_{ix} + \left(\frac{V_e + k_h}{m_a} \right) \cdot t \right)^2 \right) \cdot S_W \cdot C_L \cdot cos(\alpha_t) - W_a \right)}{m_a} \tag{8}$$

where V_v is the vertical velocity of an airplane; V_{vi} is the initial vertical velocity of an airplane; V_{ix} is the initial velocity of the airplane; V_e is the engine velocity of the airplane; k_h is the trust constant; m_a is the mass of an airplane; t is the flight time; S_w is the wing surface; C_L is the coefficient of lift; α_t is the angle of rotation; and W_a is the weight of an airplane.

4.4.3 Mathematical model of lateral velocity

According to [21–23], the resultant force L_y of the sinusoidal component of the lift is responsible for generating the speed of lateral displacement of the aircraft. The following equation is obtained:

$$V_L = V_{iy} \frac{t \cdot \left(\left(\rho \left(V_{ix} + \left(\frac{V_e + k_h}{m_a} \right) \cdot t \right)^2 \right) \cdot S_W \cdot C_L \cdot sin(\alpha_t) \right)}{m_a} \tag{9}$$

where V_L is the lateral velocity of an airplane; V_{iy} is the initial velocity on the y-axis of an airplane; V_{ix} is the initial velocity on the x-axis of the airplane; V_e is the engine velocity of an airplane; k_h is the trust constant; m_a is the mass of an airplane; t is the flight time; S_w is the wing surface; C_L is the lift coefficient; and α_t is the angle of rotation.

4.5 Implementation and testing

The CPS complied with a development process for the flight simulator based on the XP cycle that performs iterative and incremental tasks [24, 25]. According to the XP methodology, the work team completed incremental delivery of the software products, based on the following iterations: (1) computational iteration—in this iteration, mathematical models were developed, aerodynamics in physics, management of threads and the design of delegates instance to control the airplane of the virtual world; (2) communication iteration—in this iteration, the processing and transfer of data referring to roll, pitch, and yaw was performed using text files, sockets and a local network; and (3) control iteration—in this iteration, the control system of the Stewart platform was programmed to reproduce the movements of the flight simulator. A multidisciplinary approach supported by the system design engineering allowed the integration of all the modules of the complex dynamic system for its correct operation. **Figure 7** shows the iterations of the CPS based on the XP methodology, applying the model proposed by Drake et al. [26].

Unitary tests were performed on the cyber-physical system that included the computational, communication and control subsystems, both at the software level and at the hardware level. Acceptance tests were also carried out with a group of 40 aircraft pilots, to evaluate the proper functioning of the software application at the end of each iterations. **Figure 8** shows the graphical user interface of the flight simulator developed with the Unity 3D framework, where it is possible to observe the cockpit of a Cessna 172 aircraft operated by a joystick, where the plane is flying over the city of Manta in Ecuador, where the main military aviation schools of the country are located.

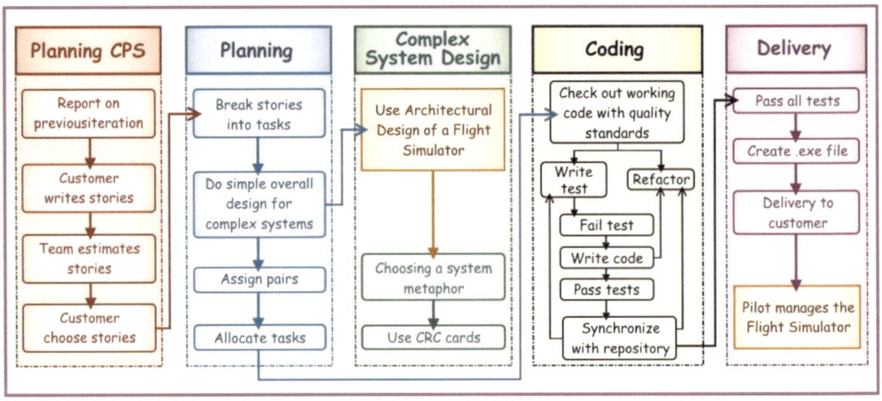

Figure 7.
The Xtreme programming CPS iterations.

Figure 8.
Graphical user interface of the flight simulator.

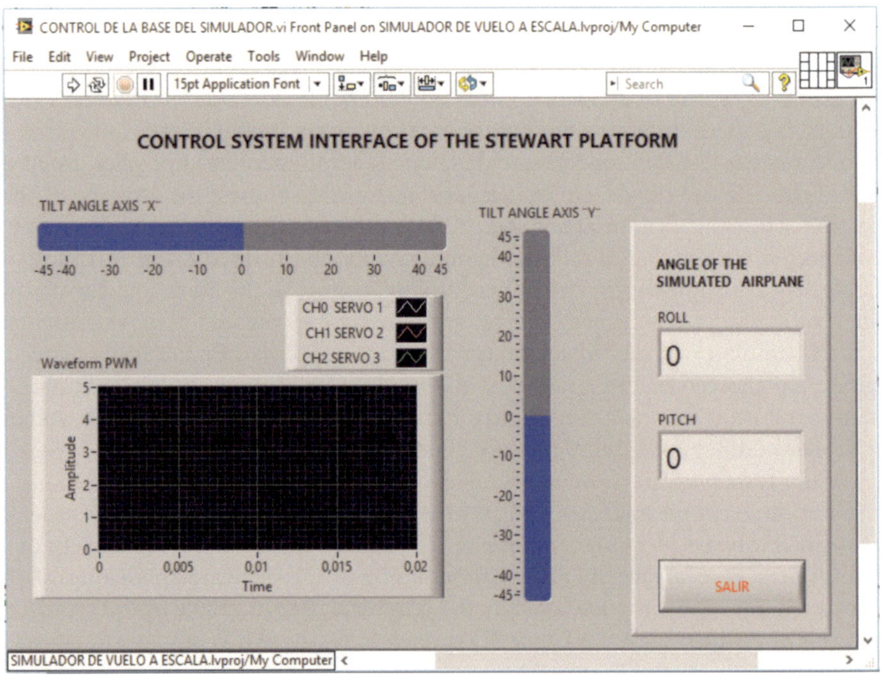

Figure 9.
Control front panel of the Stewart platform.

Figure 9 shows the control front panel of the Stewart platform, where the behavior of the input and output values of the control system developed in LabVIEW can be seen, which reproduces the movements of the flight simulator by integrating a network of two computers that use communication threads with sockets and flat text files, where the information of the three angles of maneuverability of the airplane are found, including: the direction (heading or yaw), elevation (pitch), and angle of bank (roll).

For a proof of concept related to the operation of the flight simulator with the Stewart platform at scale, please access the following links available at: https://youtu.be/pyXP5FlyJYU, https://youtu.be/13d4mFDglmM.

5. Experimental results and discussion

A total population of 40 pilots belonging to the aviation schools of the Armed Forces located in the city of Manta in Ecuador have been chosen to use, test and fly in the developed flight simulator. Of the 40 pilots, 15 correspond to the area of instructors and 25 correspond to the area of students of the aviation schools. After testing the constructed flight simulator, we proceeded to execute the statistical processing, for which we compared six basic characteristics of a flight simulator such as: (a) maneuverability capabilities; (b) motion detection (roll, pitch, yaw); (c) change of plane on the stage; (d) aerodynamic performance; (e) interaction with the virtual simulator; (f) control of the pilot board. These characteristics have been compared with the Microsoft Flight Simulator commercial simulator, obtaining the results indicated in **Figure 10**.

Figure 10, documents the results obtained from a sample of 40 pilots, as indicated above. The reference point has been executed in such a way that a pilot has been tested only in the environment of the flight simulator, being saved of expressing any comment of the experience of testing the simulator with the companions of the aviation schools. Consequently, each evaluation performed by each instructor or apprentice pilot has been free of mutual influence, which leads that the collected data have been an adequate indicator of the perception of the pilots of the flight simulator as a perceptive experience.

In this version of the flight simulator, the lowest value in the characteristics of the simulator corresponds to the control of the pilot's board, since the design and development of three of the six basic flight instruments is almost complete: (a) altimeter; (b) airspeed indicator; (c) vertical speed indicator; (d) attitude indicator; (e) heading indicator; (f) turn indicator. Actually, the simulator is 90% ready and

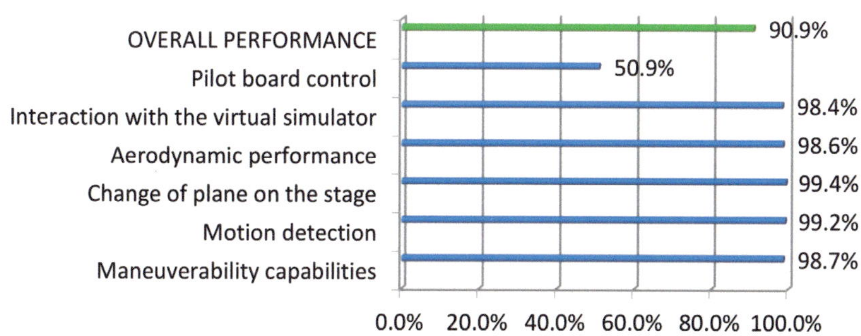

Figure 10.
Average scores of the characteristics of the selected reference points applied for the tests of the flight simulator pilots.

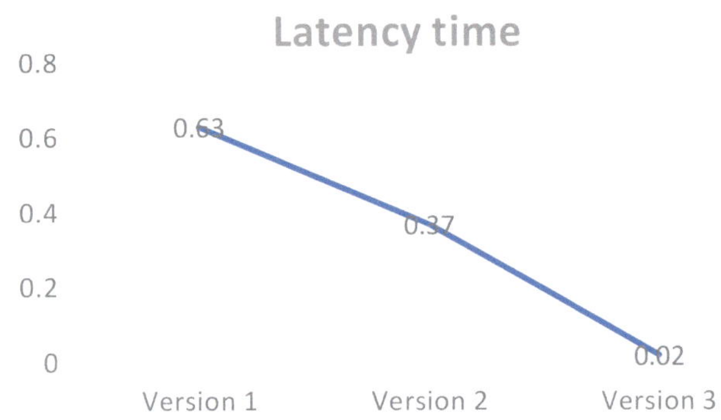

Figure 11.
Mean scores for latency time for all three versions.

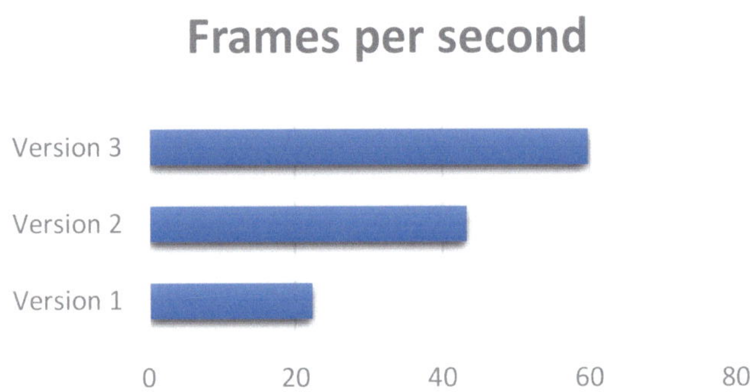

Figure 12.
Mean scores for frames per second for all three versions.

works with a joystick, while it is expected to complete the research project with a rudder, pedals and the power lever connected to the simulator through USB ports.

In addition, three experiments have been performed integrating the algorithms of the mathematical models developed with the Unity 3D game engine with the control algorithms of the Stewart platform of three degrees of freedom developed in LabVIEW, obtaining a communication algorithm that receives the variables of transformation of axes by rotation of the point P (w, x, y, z). Therefore, the Stewart's platform has been able to reproduce the movements of the simulator software, obtains the roll, pitch and yaw values. The evolution of the communication algorithm to reproduce the movement of the Stewart platform in real time with the movements that are generated in the simulator software has been as follows (see **Figures 11** and **12**):

5.1 Version 1

This algorithm contains the movement control variables that were initially written separately and then saved in a text file, in such a way that a latency time of 0.63 s was obtained with 22.3 frames per second.

5.2 Version 2

This algorithm contains the movement control variables that can be written using the Update () function, in such a way that a latency time of 0.37 s was obtained with 43.28 frames per second.

5.3 Version 3

This algorithm contains the movement control variables that can be sent by threads with the Update () function, in such a way that a latency time of 0.02 s was obtained with 59.8 frames per second.

The results obtained from the different versions of the flight simulator are shown in **Figures 11** and **12**.

Based on these obtained results, we have been able to document that the latency time has been reduced from 0.63 to 0.02 s, thus achieving data transmission of almost 60 frames per second. This allowed the reproduction in real time of the movements of the Stewart platform with the movements of the flight simulator software.

6. Conclusions and future work

The objective of this study was to design, develop and implement a flight simulator as a Cyber-Physical system with a Stewart platform at scale with three degrees of freedom. It has been fulfilled principles of System Design Engineering such as the V-Model for Mechatronics and Cyber-Physical Systems; and Software Engineering techniques such as the XP method for the process of design and development of the computer system in terms of communicational, computational and control modules, ensuring the quality of the software. In addition, mathematical models were applied for the calculations of the frontal, vertical and lateral velocities. These formulas were developed and then implemented with the Unity 3D game engine for the Cessna 172 aircraft, in a network of two computers that communicated with each other through flat files consumed by the LabVIEW libraries that allowed reproducing the movements of the flight simulator software application on the Stewart platform scale. This simulator has been created to improve skills and abilities of flight and space disorientation in the training of military pilots of war and combat aircraft. The validation of the proposed solution was made at instructor and student level with several pilots of the aviation schools of the Armed Forces of Ecuador, who have previously been trained in the handling of flight simulation software known as the Microsoft Flight Simulator. The results indicated that this flight simulator supports the development of aircraft control skills and abilities, leading to an increase in its maneuverability and flight capabilities.

As future works, we plan to raise the development and implementation of flight simulators and low-cost spatial disorientation that allow reproducing the movements of the simulation software on a Stewart platform with four degrees of freedom composed of three pistons, electro-valves, microprocessors and other electromechanical elements to reproduce the movement of the roll, pitch, and yaw, as well as to generate at least two or three gravities through the left and right rotational movement of the Stewart platform to disorient the pilot.

Acknowledgements

This study has been partially funded by the Universidad de las Fuerzas Armadas ESPE in Sangolquí, Ecuador within the research project entitled “Construction of a spatial disorientation simulator for contribute to the aviation safety and training of the pilots of Army Forces FF-AA.”

Conflict of interest

The authors declare no have conflict of interest.

Author details

César Villacís[1,2], Walter Fuertes[1,2*], Luis Escobar[1,2], Fabián Romero[1,2] and Santiago Chamorro[1,2]

1 Computer Sciences Department, University of the Army Forces-ESPE, Sangolquí, Ecuador

2 Mechanical and Energy Department, University of the Army Forces-ESPE, Sangolquí, Ecuador

*Address all correspondence to: wmfuertes@espe.edu.ec

References

[1] Guerineau B, Bricogne M, Durupt A, Rivest L. Mechatronics vs. cyber physical systems: Towards a conceptual framework for a suitable design methodology. In: 2016 11th France-Japan & 9th Europe-Asia Congress on Mechatronics (MECATRONICS)/17th International Conference on Research and Education in Mechatronics (REM); Compiègne, France; 2016, pp. 314-320

[2] Valis D, Vintr Z, Soares C, Zio E. Dependability of mechatronics systems in military vehicle design. Safety and Reliability for Managing Risk–Proceedings of the European Safety and Reliability Conference 2006 (ESREL). 2006

[3] Mezyk A, Switonski E, Kciuk S, Klein W. Modelling and investigation of dynamic parameters of tracked vehicles. Mechanics and Mechanical Engineering. 2012;**15**(4):115-130

[4] Rizzoni G, et al. Modeling, simulation, and concept design for hybrid-electric medium-size military trucks Defense and Security. Orlando, FL; 2005. pp. 1-12

[5] Escobar L, et al. Design and implementation of complex systems using mechatronics and cyber-physical systems approaches. In: 2017 IEEE International Conference on Mechatronics and Automation (ICMA); Takamatsu, Japan; 2017. pp. 147-154. DOI: 10.1109/ICMA.2017.8015804

[6] Plateaux R, Penas O, Choley J, Mhenni F, Hammadi M, Louni F. Evolution from mechatronics to cyber physical systems: An educational point of view. In: 2016 11th France-Japan & 9th Europe-Asia Congress on Mechatronics (MECATRONICS)/17th International Conference on Research and Education in Mechatronics (REM); Compiegne; 2016, pp. 360-366. DOI: 10.1109/MECATRONICS.2016.7547169

[7] Hager G, Wellein G. Introduction to High Performance Computing for Scientists and Engineers. USA: Taylor and Francis Group, LLC, CRC Press; 2010. ISBN 978-1-4398-1192-4

[8] Derler P, Lee EA, Vincentelli AS. Modeling cyber–physical systems. Proceedings of the IEEE. 2012;**100**(1): 13-28. DOI: 10.1109/JPROC.2011.2160929

[9] Rajkumar R, Lee I, Sha L, Stankovic J. Cyber-physical systems: The next computing revolution. In: Proceedings of the 47th Design Automation Conference on—DAC'10; Anaheim, California; 2010. pp. 731

[10] Lee EA. Cyber-physical systems-are computing foundations adequate. Position paper for NSF workshop on cyber-physical systems: Research motivation, techniques and roadmap. Vol. 2. Citeseer, 2006

[11] Heynen M. Cluster Computing: Distributed Computing Architecture. 1st ed. CreateSpace Independent Publishing Platform. ISBN: 978-1541018563. USA; 2016

[12] Vepa R. Flight Dynamics, Simulation, and Control: For Rigid and Flexible Aircraft. 1st ed. First Published, CRC Press; 2014, 696 pages. eBook ISBN: 9780429101724. DOI: https://doi.org/10.1201/b17346

[13] Hunt KH. Structural kinematics of in-parallel-actuated robot-arms. Journal of Mechanisms, Transmissions, and Automation in Design. 1983;**105**(4):705-712. 8 pages. DOI:10.1115/1.3258540

[14] Alexandrov VV et al. Introduction to Control of Dynamic Systems. Ediciones BUAP. 2009. p. 239. ISBN 978-607-487-095-4

[15] Matlalcuatzi E et al. Diseño del simulador dinámico para pilotos como

un sistema biomecatrónico. In: Memorias del XVI Congreso Latinoamericano de Control Automático, CLCA 2014; Octubre 14–17; Cancún, Quintana Roo, México

[16] Rekdalsbakken W. Design and application of a motion platform in three degrees of freedom. In: Proceedings of SIMS 2005, the 46th Conference on Simulation and Modelling; NO-7005 Trondheim: Tapir Academic Press. pp. 269-279

[17] Rekdalsbakken W. Design and application of a motion platform for a high-speed craft simulator. In: ICM 2006, IEEE 3rd International Conference on Mechatronics. DOI: 10.1109/ICMECH.2006.252493

[18] Villacís C, Navarrete M, Rodríguez I, Romero F, Escobar L, Fuertes W, et al. Real-time flight simulator construction with a network for training pilots using mechatronics and cyber-physical system approaches. 2017 IEEE International Conference on Power, Control, Signals and Instrumentation Engineering (ICPCSI), Chennai, India; 2017, pp. 238-247. DOI: 10.1109/ICPCSI.2017.8392169

[19] Zheng S, He J, Jin J, Han J. DDS based high fidelity flight simulator. In: WASE International Conference on Information Engineering; IEEE; 2009. DOI 10.1109/ICEI.2009.61

[20] Xinren W, Rongzhen J, Xiaoyuan P, Qin F. Flight Real-Time Simulation System and Technology. Beijing: Press of Beihang University; 2003

[21] Merk RJ, Roessingh JJM. Assessing behaviour of cognitive agents in a flight simulator with fighter pilots. In: 2016 IEEE International Conference on Systems, Man, and Cybernetics (SMC); IEEE; 2016

[22] Boril J, Leuchter J, Smrz V, Blasch E. Aviation simulation training in the Czech air force. In: 2015 IEEE/AIAA 34th Digital Avionics Systems Conference (DASC); IEEE; September 2015. pp. 9A2-1

[23] Villacís C et al. Mathematical models applied in the design of a flight simulator for military training. In: Rocha Á, Guarda T, editors. Developments and Advances in Defense and Security. In: MICRADS 2018. Smart Innovation, Systems and Technologies. vol. 94. Cham: Springer; 2018, pp. 43-57

[24] Beck K. Extreme Programming Explained: Embrace Change. 1st ed. Addison-Wesley Professional. USA: Pearson Education Inc. Nov 17th, 2004, 189 pages. ISBN: 0321-27865-8

[25] Pressman RS. Ingeniería de Software. Un Enfoque práctico. Sexta edición ed. México DF, México: Editorial McGraw-Hill; 2010. ISBN: 9701054733

[26] Drake P, Kerr N, Arbor A. Developing a computer strategy game in an undergraduate course in software development using extreme programming. In: Eighth Annual North-Western Regional Conference; 2006

Chapter 5

Detecting Underground Military Structures Using Field Spectroscopy

George Melillos, Kyriacos Themistocleous, Athos Agapiou, Silas Michaelides and Diofantos Hadjimitsis

Abstract

Satellite remote sensing is considered as an increasingly important technology for detecting underground structures. It can be applied to a wide range of applications, as shown by various researchers. However, there is a great need to integrate information from a variety of sources, sent at different times and of different qualities using remote sensing tools. A SVC-HR1024 field spectroradiometer could be used, and in-band reflectance's are determined for medium- and high-resolution satellite sensors, including Landsat. Areas covered by natural soil where underground structures are present or absent can easily be detected, as a result of the change in the spectral signature of the vegetation throughout the phenological stages; in this respect, vegetation indices (VIs) such as the normalized difference vegetation index (NDVI), simple ratio (SR), and enhanced vegetation index (EVI) may be used for this purpose. Notably, the SR vegetation index is useful for determining areas where military underground structures are present.

Keywords: spectroscopy, military underground structures, vegetation indices, Landsat

1. Introduction

For decades, research on the detection of buried targets has led to the development of a variety of techniques for identifying buried structures [1, 2]. These techniques use a variety of geophysical instruments [3–6] that involve the use of ground-penetrating radar (GPR). Ground-penetrating radar is a sensitive technique for detecting even small changes in the subsurface dielectric constant. Consequently, the images generated by GPR systems contain a great amount of detail, much of it either unwanted or unnecessary for purposes of surveying for underground objects. A major difficulty, therefore, in using GPR for locating an underground structure concerns the present inability in the art to correctly distinguish return signals reflected by an underground object of interest from all of the signals generated by other subsurface features; the latter signals are collectively referred to as clutter [7].

Nowadays, a lot of attention is being paid to the development of new methods and instrumentation for the detection of buried targets. The detection of military underground structures is a major concern for military and national security agencies, as this is evident from the large budget [8] allocated for the detection and monitoring underground structures. National security agencies use human intelligence (HUMINT) as one of the currently used information collection methods. HUMINT refers to the collection of information by a trained HUMINT collector (military occupational specialty) [8], from people and their associated documents and media sources for the identification of elements, intentions, composition, strength, dispositions, tactics, equipment, personnel, and capabilities. Additionally, technology such as imagery intelligence (IMINT) can also be used for gathering information via satellite and aerial photography. Remote sensing techniques are quick, are easily manageable, and involve a wide variety of techniques where valuable information can be accessed remotely [9, 10].

Buried underground structures are difficult to detect, especially when they are fully covered by soil [11]. It is possible to detect such military underground structures by means of satellite images and aerial photographs. The concern about underground facilities (or "hard and buried" targets) is evident from the establishment of several purpose-dedicated components within various intelligence and defense agencies [12].

Underground structures such as military constructions and archaeological remains can affect their surrounding landscapes in different ways, such as localized soil moisture content and drainage rates [13], soil composition, and vegetation vigor [7]. Vegetation vigor could be observed on the ground as a crop mark, a spot which can be used to indicate the presence of underground structures [14]. Crop marks can be formed both as negative marks above concrete foundations and as positive marks above the damper and more nutritious soil of buried pits and ditches [14].

During the last decade, the improvement of sensor characteristics, such as higher spatial resolution and hyperspectral data, as well as the technological achievements in space technology, offers new opportunities for future applications [15].

Additionally, in some cases, researchers seek not to find the target itself but rather to identify symptoms related to the topography (relief), crop characteristics (crop marks), soil characteristics (soil marks), or even changes in snow cover (snow marks). For instance, archaeological structures buried beneath the soil (i.e., still un-excavated sites) can be detected through remote sensing images as stressed vegetation (crop marks) which can be used as a proxy for the buried archaeological relics. Crop marks may be formed in areas where vegetation grows over near-surface archaeological remains. These features modify the moisture retention compared to the rest of the crop coverage of an area. Depending on the type of the feature, crop vigor may be enhanced or reduced by buried archaeological features [16].

In comparing the two different kinds of marks, the positive crop marks are normally taller with darker green and healthy foliage than the negative crop marks, while negative crop marks tend to be paler green with lighter-colored appearance when monitored from the air [17]. Indeed, spectral remote sensing is widely used in several occasions for the detection of underground structures, such as agricultural remains [18].

In addition, spectral remote sensing for the detection of underground military structures is considered to be very precise in detecting subsurface remains. Different geophysical processing techniques and equipment, such as GPR, magnetometer, and resistivity, are usually integrated to maximize the success rate of uncovering underground remains [19–22]. Moreover, the use of unmanned aerial vehicle (UAV, popularly known as a drone) for environmental remote sensing purposes has increased in recent years. Although the military has used UAVs for defense applications for decades, the scientific environmental sector increasingly takes advantage of the application of UAVs [23].

Also, this chapter investigates the possibility of applying satellite data using Landsat 8 sensor and comparing it with field data in an effort to distinguish between buried structure area (existence of underground structure) and vegetated area (lack of underground structure).

It is noted that in the literature, there is a gap in the monitoring of vegetation over military underground and ground structures throughout the plants' phenological development cycle; this paper aspires to contribute to the filling-in of this gap. Indeed, this chapter aims at presenting the results obtained from ground spectroradiometric campaigns, using an SVC-HR1024 field spectroradiometer, carried out in a specific area in Cyprus. For in situ observations, field spectroradiometric data were collected and analyzed to identify (known) underground structures using the spectral profile of the vegetated surface over the underground target and the surrounding area. Crop marks demonstrate the variations between the presence and the absence of military underground structures. The in situ measurements were resampled to the Landsat 8 sensor using the appropriate relative spectral response (RSR) filters.

2. Materials and methods

This study proposes a methodology for detecting underground targets using remote sensing techniques. The basis of this methodology is the combination of the study of the vegetation phenology as a proxy for buried underground structures of significance to defense. Data acquisitions were used to identify any variations between the area over an underlying structure and over a reference area.

For this study, certain assumptions have been adopted. In the case of this project, phenological field observations were conducted in two test sites from 2016 to 2017 to determine the dates of completion of different phenological phases. For actual defense purposes, the characteristics of the area of interest are often not known. Furthermore, the cultivation of barley in the area is for investigative purposes and part of the experimental work for studying the impact of underlying structures on vegetation. Under real scenarios, different types of vegetation (if any) will be present.

2.1 Study area

The proposed methodology has been applied in Cyprus over a specific geographical area. The area is situated on a hill which provides clear viewing from airborne and spaceborne platforms, making the area ideal for remote sensing applications (**Figure 1**, left). Also, it is located within a fenced, abandoned military area (due to security and confidentiality issues, the specific area cannot be reported herein). The soil type of the area is leptosol which contains small amounts of gravel and with a very shallow depth.

Figure 1.
Overview of the study area (left); an example of a military bunker (right).

Figure 1 (right) shows a military storage bunker similar to the one that is in the focus of this research. The horizontal dimensions of the underground structure are 13 m × 5 m; it is a concrete storage bunker, located ~2 m below the ground surface.

2.2 Methodology

Spectral data are increasingly incorporated into process-based models of the Earth's surface and the atmosphere. The area of interest was determined first by identifying plots with a high probability of buried targets. Such areas can be determined from various sources, such as on-site irregular activities, personal communication, surveys, and crop marks.

In situ measurements were taken at two test areas: (a) vegetation area covered with vegetation (barley), in the presence of an underground military structure [hereafter denoted as area (a)], and (b) vegetation area covered with vegetation (barley), in the absence of an underground military structure [hereafter denoted as area (b)].

An SVC-1024 spectroradiometer of the Spectra Vista Corporation (SVC) with a spectral range of 350–2500 nm was used to measure reflectance values. The spectral resolution of the spectroradiometer is 1.0 nm. The measurements were taken between 11:00 AM and 13:00 PM (local time), under clear and overcast skies for diffuse light to minimize any variation of the incoming solar electromagnetic radiance. In addition, a calibrated Spectralon panel (with reflectance ≈99.996%) measurement was used as a reference, while the measurements over the crops were used as a target [12].

Then, the waveband reflectance values were used to calculate three vegetation indices (VIs), the normalized difference vegetation index (NDVI), enhanced vegetation index (EVI), and simple ratio (SR), as shown in **Table 1**. The waveband reflectance's were calculated from the RSR filter of the Landsat 8 sensor. The vegetation indices were plotted and statistically cross compared between the two areas of interest, namely, the "buried military structure" and the "nonmilitary structure."

In situ measurements were taken in a grid format which is the same as the dimension of the underground structure (13 m × 5 m), over the two study areas, and systematic targets were collected at each time to compile a representative sample that is statistically reliable. Due to the very close proximity of the two sites (<20 m), the analysis was based on the following two criteria: both study areas have similar soil and climatic characteristics [28]. Area (a) is the area over the underground structure itself and the area around it. Area (b) is the reference area in the absence of an underground military structure. The measurements were also made when the underground structure was covered with the existing natural soil which was subsequently cultivated and covered with vegetation (barley) in order to study possible differences of the spectral signature of vegetation, as a result of the existence of underground structures. During the campaign, 1740 measurements were taken using the SVC-1024 field spectroradiometer, with an average reference spectral signal as given for each of the six campaigns (random sampling) in **Table 2**.

No.	Vegetation index	Equation	Reference
1.	NDVI (Normalized Difference Vegetation Index)	$(p_{NIR} - p_{RED})/(p_{NIR} + p_{RED})$	[25]
2.	EVI (Enhanced Vegetation Index)	$2.5\ (p_{NIR} - p_{RED})/(p_{NIR} + 6\ p_{RED} - 7.5\ p_{BLUE} + 1)$	[26]
3.	SR (Simple Ratio)	p_{NIR}/p_{RED}	[27]

Table 1.
Vegetation indices used in this study, where p_{NIR}, p_{RED}, and p_{BLUE} represent the atmospherically or partial-atmospherically corrected surface reflectance values of the near-infrared (NIR), red (RED), and blue (BLUE) wavelengths, respectively [24].

No.	Date	Phenological stage	Number of measurements
(a)	30-10-2016	Cultivation stage	120
(b)	11-12-2016	Tilling stage	120
(c)	23-01-2017	Flag Leaf Emerging stage	120
(d)	25-02-2017	Boot stage	460
(e)	05-03-2017	Head Emerging stage	460
(f)	16-03-2017	Flowering stage	460

Table 2.
Number of measurements in each phenological stage.

3. Results

3.1 Vegetation indices

Figures 2-4 show the results for the vegetation indices shown in **Table 1** with reference to the phenological stages. The vegetation indices were applied to the barley crop over area (a) (red line) and area (b) (blue line). The response of VIs with respect to barley growth was evaluated by contrasting the minimum to the abovementioned areas. The results show that VIs display a distinct variation corresponding to the barley development and they could be used as cultivar-independent phenological indicators. It can be observed that there is a high correlation between the results of VIs. Indeed, VIs could be used in field spectroscopy for the detection of buried structures. The use of more than one VI for the detection of crop marks is suggested in order to enhance the final results. Furthermore, it is clear from these graphs that VI values vary from one phenological stage to another. Although the same dataset was used for all these vegetation indices, each of the VIs demonstrates

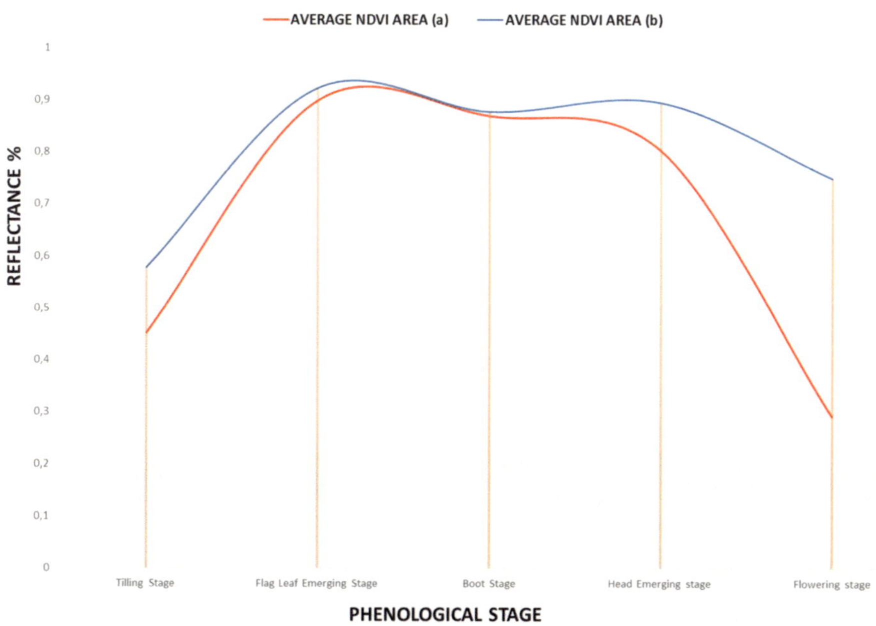

Figure 2.
Vegetation values for area (a) for (buried structure, red dots) and area (b) for (vegetated area, blue line) during phenological cycle for NDVI.

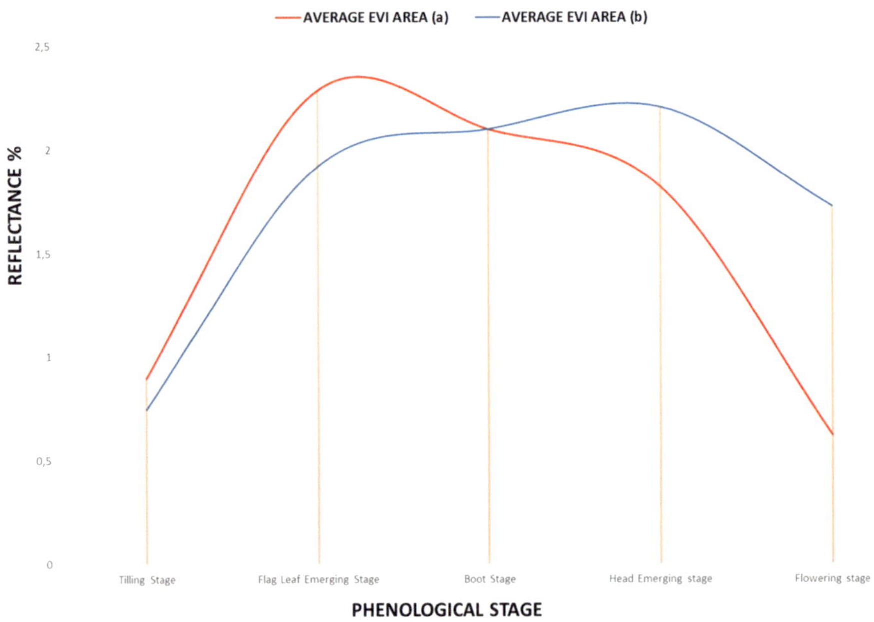

Figure 3.
Vegetation values for area (a) for (buried structure, red dots) and area (b) for (vegetated area, blue line) during phenological cycle for EVI.

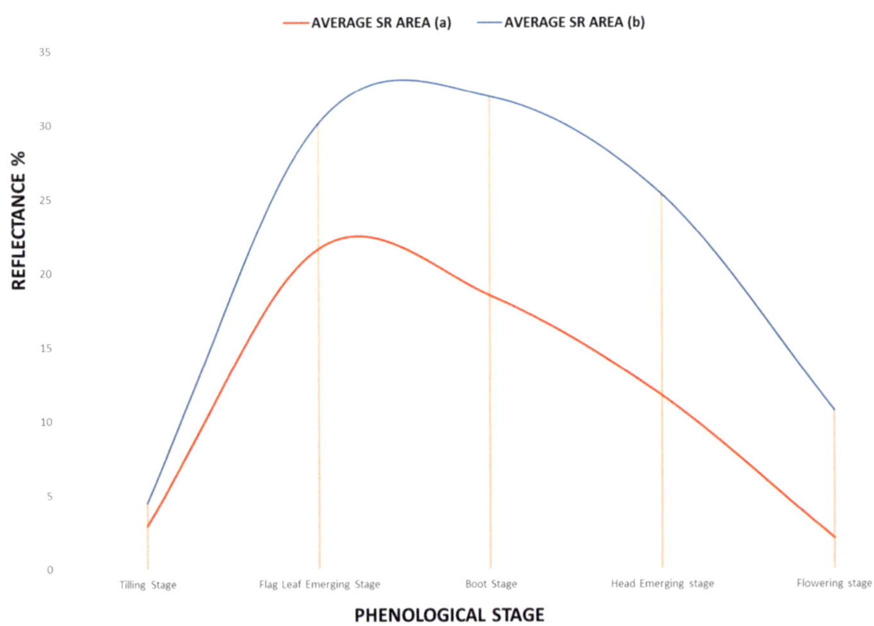

Figure 4.
Vegetation values for area (a) for (buried structure, red dots) and area (b) for (vegetated area, blue line) during phenological cycle for SR.

a different response at different phenological stages. It may be seen clearly in the Flowering stage that there is a distinction between area (a) and area (b). **Figure 2** presents a typical example of the spectral profile of area (a) and area (b) using the NDVI. There is an upward trend of area (b) (blue line) compared with area (a) (red line) in which there is a downward trend, throughout the phenological cycle. Evidently, this happens using the SR vegetation index. Moreover, it is remarkable to note that for the SR vegetation index (**Figure 3**), in test area (b) (blue line), the reflectance response is higher than test area (a) (red line), throughout the phenological cycle. Obviously, the reflectance response in test area (b) (blue line) follows a steeper upward path in the Tilling and Flag Leaf Emerging stages. In the Boot stage, the reflectance of area (b) (blue line) increases dramatically. In the Head Emerging and Flowering stages, the reflectance decreases, but there is a differentiation between the two test areas. This differentiation is not arbitrary but reinforces the diagnosis of existence/nonexistence of military underground structures. In addition, for the EVI (**Figure 4**), the reflectance response changes over time. Specifically, there is an upward trend in area (a) (red line) in the Tilling and Flag Leaf Emerging stages. In contrast, there is an upward trend in area (b) (blue line) in the Head Emerging and Flowering stages.

3.2 Image maps

Ground spectroradiometric measurements can provide the spectral response of the vegetation in detail [29]. The analysis of the spectral data shows the maps of vegetation indices (NDVI, EVI, and SR) for area (a) (**Figure 5**) and area (b) (**Figure 6**), during Flag Leaf Emerging stage. Comparing area (a) with area (b) using NDVI (**Figure 5**), it appears that area (b) obtains lower values due to the nonexistence of underground structures, while area (a) has similar vegetation but higher NDVI values due to the existence of underground structures. In addition, using the EVI (**Figure 6**), area (a) has higher values due to the existence of underground structures, while area (b) has lower values due to nonexistence of structures. Similarly, using the SR vegetation index (**Figure 6**), area (b) has clearly lower values due to the existence of underground structures, while area (a) (**Figure 5**) has higher values due to nonexistence of structures. The green color illustrates high value of indices that distinguish the existence of structures. The existence of underground structure can be clearly seen by comparing area (a) with area (b), during the Head Emerging stage. The analysis of the spectral data shows the maps of vegetation indices (NDVI, EVI, and SR) for area (a) (**Figure 7**) and area (b)

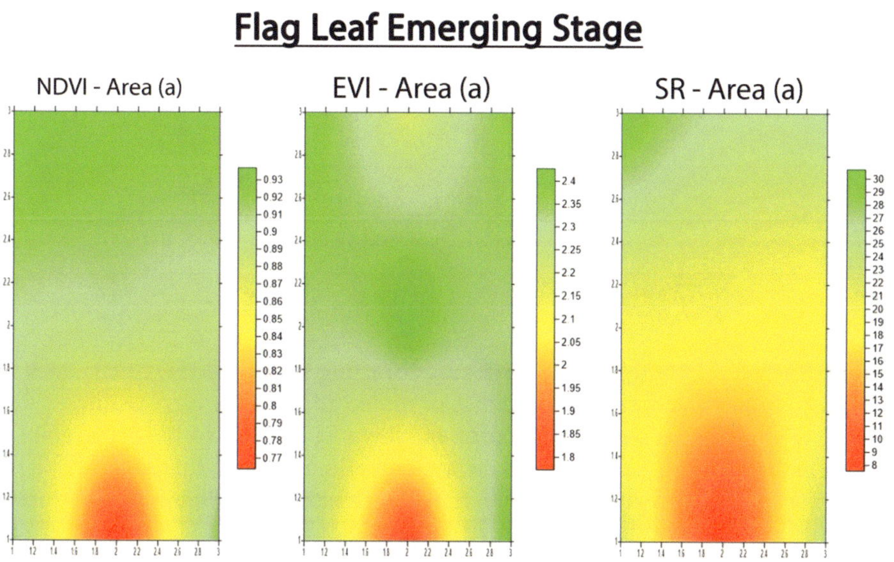

Figure 5.
Image maps showing NDVI, EVI, and SR field data for area (a) during Flag Leaf Emerging stage.

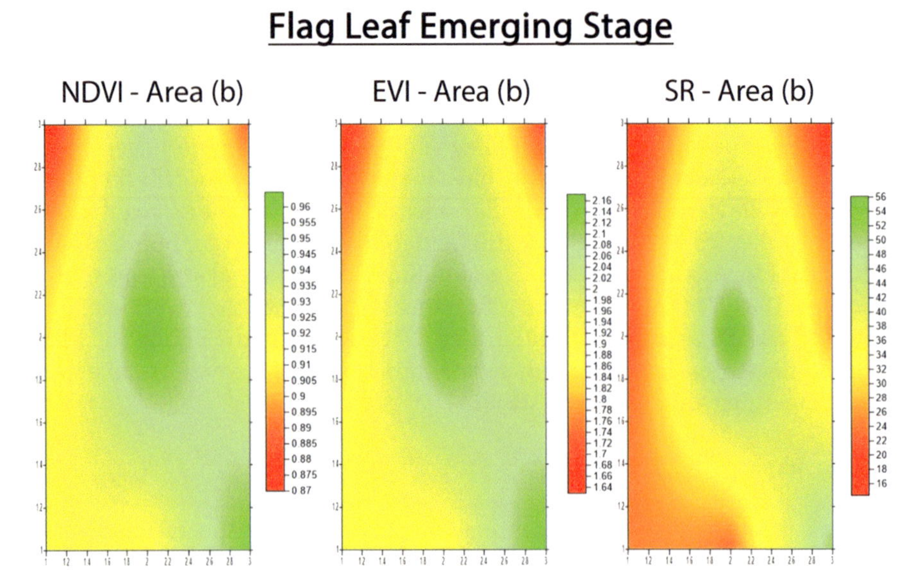

Figure 6.
Image maps showing NDVI, EVI, and SR field data for area (b) during Flag Leaf Emerging stage.

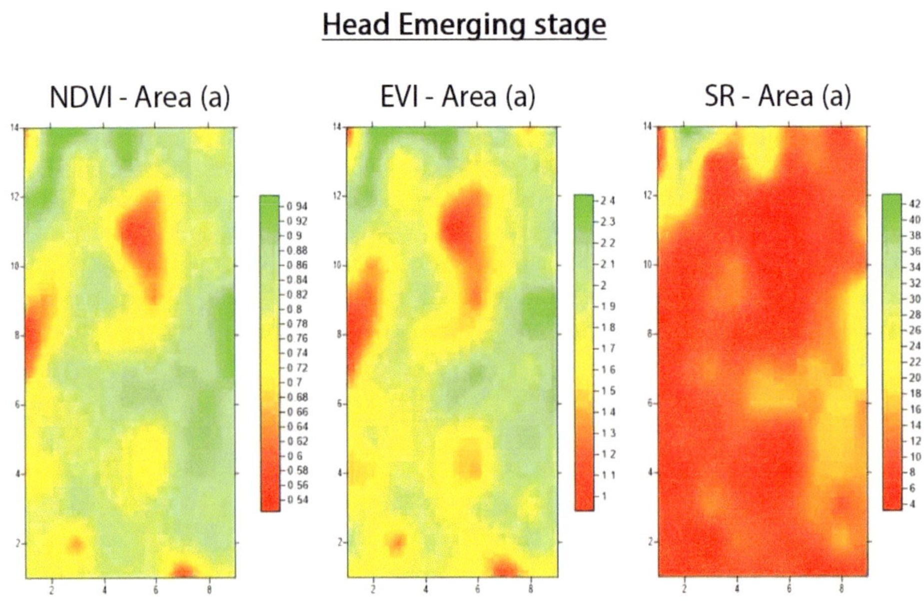

Figure 7.
Image maps showing NDVI, EVI, and SR field data for area (a) during Head Emerging Stage.

(**Figure 8**). In comparing area (a) with area (b) using the NDVI (**Figure 8**), area (b) has higher values due to the nonexistence of underground structures, while area (a) (**Figure 7**) has similar vegetation but lower NDVI values due to the existence of underground structures. Likewise, using the EVI, area (a) (**Figure 7**) has lower values due to the existence of underground structures, while area (b) (**Figure 8**) has higher values due to nonexistence of structures.

Using the SR vegetation index (**Figure 7**), area (a) tends to exhibit a difference with respect to area (b) (**Figure 8**). More specifically, in target area (a) (**Figure 7**), the reflectance response is lower than area (b) (**Figure 8**), which indicates that the

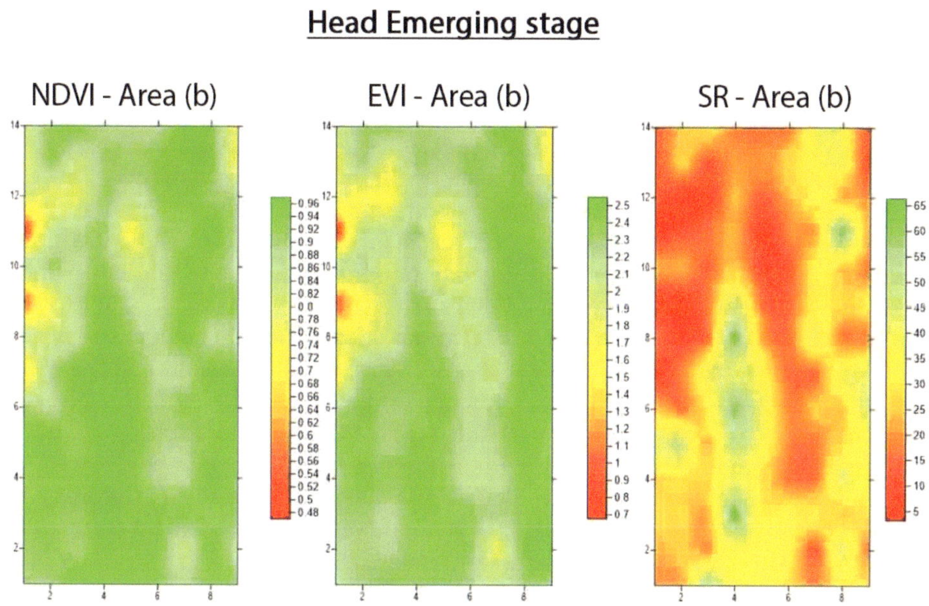

Figure 8.
Image maps showing NDVI, EVI, and SR field data for area (b) during Head Emerging Stage.

resulting differences reinforce the existence/nonexistence of underground structures. It can be argued that soil also contributed to the reflectance measurements. The variations between the two cases, namely, in the presence and in the absence of military underground structures, can result in better interpretations of images for the detection and identification of crop marks.

4. Conclusions

Field spectroscopy can support satellite remote sensing studies for monitoring systematically critical areas of interest including the detection of underground bunkers.

The application of remote sensing in defense and security merges the technological improvements of remote sensing sensors with military needs to improve the quality of information retrieved from remote sensing data. Indeed, decision-making authorities can benefit from such efficient space imaging technology of underground targets.

The advantages of using vegetation indices as proxy variables for intercalibration among existing sensors are the low sensitivity to the uncertainties in atmospheric correction and the variation in the satellite viewing angle [30]. As shown in this paper, vegetation indices can corroborate areas of possible military underground structure.

In comparing the two areas, the spatial distributions of VIs exhibit no table differences (**Figures** 7 and **8**). This is clear in **Figure 8**, where image maps illustrate the differences between the two areas using NDVI, EVI and SR Vegetation Indices. Mostly SR vegetation index is used for defining areas where military underground structures are present.

Consequently, the near-infrared (NIR) band of Landsat 8 sensor could be useful to identify the underground structure. It is apparent that the waveband analysis of the Landsat 8 sensor distinguishes between the two study sites. Monitoring variations of the NIR spectrum during the life cycle of vegetation is a key parameter for the field spectroscopy for the detection of military underground structures using remote sensing techniques.

In this chapter, it was demonstrated how remote sensing can be exploited as a monitoring and decision-making tool by any agency in tackling military and security issues related to the presence of underground military structures. Field spectroscopy measurements were used to detect underground military structures through variations in vegetation indices. Indeed, vegetation indices can be used to develop a suitable vegetation index for detecting military underground structures.

Areas covered by natural soil where underground structures are present or absent can easily be detected as a result of the change in the spectral signature of the overlying vegetation; in this respect, vegetation indices, such as the NDVI, SR, and EVI, may be used for this purpose.

It is recommended to collect field spectroradiometric measurements to other types of military underground structures to evaluate the above results and the satellites' spectral sensitivity. The development of a standard model/methodology framework to be produced through the stages of the study for locating military underground structures is an innovation in military operations research. Additionally, an unmanned aerial vehicle (UAV) may be used to survey the area with visible and near-infrared cameras to generate vegetation indices for comparison to the in situ spectroradiometric measurements [31].

Acknowledgements

Acknowledgments are due to the Remote Sensing Laboratory of the Department of Civil Engineering and Geomatics, "ERATOSTHENES Research Centre" at the Cyprus University of Technology, for its continuous support (http://www.cyprusremotesensing.com).

Conflict of interest

The authors declare no conflict of interest.

Author details

George Melillos*, Kyriacos Themistocleous, Athos Agapiou, Silas Michaelides and Diofantos Hadjimitsis
Department of Civil Engineering and Geomatics, Faculty of Engineering and Technology, Cyprus University of Technology, Lemesos, Cyprus

*Address all correspondence to: gn.melillos@edu.cut.ac.cy

References

[1] Piper JE, Lim R, Thorsos EI, Williams KL. Buried sphere detection using a synthetic aperture sonar. IEEE Journal of Oceanic Engineering. 2009;**34**(4): 485-494. DOI: 10.1109/joe.2009.2030971

[2] Zhang Y, Liao X, Carin L. Detection of buried targets via active selection of labeled data: Application to sensing subsurface UXO. IEEE Transactions on Geoscience and Remote Sensing. 2004;**42**(11):2535-2543. DOI: 10.1109/tgrs.2004.836270

[3] Apparao A, Rao TG, Sastry RS, Sarma VS. Depth of detection of buried conductive targets with different electrode arrays in resistivity prospecting. Geophysical Prospecting. 1992;**40**(7):749-760. DOI: 10.1111/j.1365-2478.1992.tb00550.x

[4] Kelly RE. Underground structure detection by surface magnetic gradient measurements. In: Subsurface and Surface Sensing Technologies and Applications III. Vol. 4491. Los Alamos National Laboratory: International Society for Optics and Photonics; 2001. pp. 367-375. DOI: 10.1117/12.450182

[5] Mahrer KD, List DF. Radio frequency electromagnetic tunnel detection and delineation at the Otay Mesa site. Geophysics. 1995;**60**(2):413-422. DOI: 10.1190/1.1443778

[6] Stolarczyk LG. Gradiometer antennas for detection of tunnels by scattered electromagnetic waves. Combat Engineering Directorate, U.S. Army Belvoir Research, Development, and Engineering Center. In: Fourth Tunnel Detection Symposium on Subsurface Exploration Technology Proceedings. Golden, Colorado; 1993. pp. 26-29

[7] Stump GS, Allen CT. U.S. Patent No. 5,819,859. Washington, DC: U.S. Patent and Trademark Office; 1998

[8] Department of the Army. Human Intelligence Collector Operations: Field Manual No. 2-22. 3. Washington, DC: Mundus Publishing; 2006

[9] Canada Centre for Remote Sensing. Fundamentals of Remote Sensing [Internet]. 2016. Available from: http://www.nrcan.gc.ca/node/9309 [Accessed: 19 November 2017]

[10] Garaba SP, Zielinski O. An assessment of water quality monitoring tools in an estuarine system. Remote Sensing Applications: Society and Environment. 2015;**2**:1-10. DOI: 10.1016/j.rsase.2015.09.001

[11] Hoepffner N, Brown CA, Platt T, Stuart V. Why Ocean Colour? The Societal Benefits of Ocean-Colour Technology. Dartmouth, NS, Canada: International Ocean Colour Coordinating Group (IOCCG); 2008

[12] Papadavid G. Estimating Evapotranspiration for Annual Crops in Cyprus Using Remote Sensing [PhD thesis]. Department of Civil Engineering and Geomatics, Faculty of Engineering and Technology, Cyprus University of Technology; 2011

[13] KMC. PriddyMineries: A Walk Through History. [Internet]. 2008. Available from: http://mikek.org.uk/friendsrlm/03_priddy_mineries.pdf [Accessed: 19 March 2016]

[14] Lasaponara R, Masini N. Identification of archaeological buried remains based on the normalized difference vegetation index (NDVI) from quickbird satellite data. IEEE Geoscience and Remote Sensing Letters. 2006;**3**(3):325-328. DOI: 10.1109/lgrs.2006.871747

[15] Giardino MJ. A history of NASA remote sensing contributions to archaeology. Journal of Archaeological

Science. 2011;**38**(9):2003-2009. DOI: 10.1016/j.jas.2010.09.017

[16] Winton H, Horne P. National archives for national survey programmes: NMP and the English heritage aerial photograph collection. Landscapes through the lens. Aerial photographs and historic environment. Aerial Archaeology Research Group. 2010;**2**:7-18

[17] Hamilton D. Remote sensing in archaeology: An explicitly North American perspective. Edited by Jay K. Johnson. University of Alabama Press, Tuscaloosa, Alabama, 2006. No. of pages: 344. ISBN 0-8173-5343-7. Archaeological Prospection. 2007;**14**(2):149-150

[18] Agapiou A, Hadjimitsis DG, Alexakis DD, Papadavid G. Examining the phenological cycle of barley (*Hordeum vulgare*) using satellite and in situ spectroradiometer measurements for the detection of buried archaeological remains. GIScience & Remote Sensing. 2012;**49**(6):854-872. DOI: 10.2747/1548-1603.49.6.854

[19] Domínguez RE, Bandy WL, Gutiérrez CA, Ramírez JO. Geophysical-Archaeological Survey in Lake Tequesquitengo, Morelos, Mexico. Geofísica Internacional. 2013;**52**(3):261-275. DOI: 10.1016/s0016-7169(13)71476-4

[20] Sarris A, Galaty ML, Yerkes RW, Parkinson WA, Gyucha A, Billingsley DM, et al. Geophysical prospection and soil chemistry at the Early Copper Age settlement of Veszto-Bikeri, Southeastern Hungary. Journal of Archaeological ScienceJournal of Archaeological Science. 2004;**31**(7): 927-939. DOI: 10.1016/j.jas.2003.12.007

[21] Sarris A, Papadopoulos N, Agapiou A, Salvi MC, Hadjimitsis DG, Parkinson WA, et al. Integration of geophysical surveys, ground hyperspectral measurements, aerial and satellite imagery for archaeological prospection of prehistoric sites: The case study of Vésztő-Mágor Tell, Hungary. Journal of Archaeological Science. 2013;**40**(3):1454-1470. DOI: 10.1016/j.jas.2012.11.001

[22] Novo A, Solla M, Fenollós JL, Lorenzo H. Searching for the remains of an Early Bronze Age city at Tell Qubr Abu al-'Atiq (Syria) through archaeological investigations and GPR imaging. Journal of Cultural Heritage. 2014;**15**(5):575-579. DOI: 10.1016/j.culher.2013.10.006

[23] d'Oleire-Oltmanns S, Marzolff I, Peter K, Ries J. Unmanned aerial vehicle (UAV) for monitoring soil erosion in Morocco. Remote Sensing. 2012;**4**(11):3390-3416. DOI: 10.3390/rs4113390

[24] Melillos G, Themistocleous K, Papadavid G, Agapiou A, Prodromou M, Michaelides S, et al. Integrated use of field spectroscopy and satellite remote sensing for defence and security applications in Cyprus. In: Fourth International Conference on Remote Sensing and Geoinformation of the Environment (RSCy2016). Vol. 9688. Paphos, Cyprus: International Society for Optics and Photonics; 2016. p. 96880F. DOI: 10.1117/12.2241207

[25] Rouse JW Jr, Haas RH, Schell JA, Deering DW. Monitoring the Vernal Advancement and Retrogradation (Green Wave Effect) of Natural Vegetation. Greenbelt, MD, USA: NASA/GSFC Final Report; NASA; 1974

[26] Huete AR, Liu HQ, Batchily KV, Van Leeuwen WJ. A comparison of vegetation indices over a global set of TM images for EOS-MODIS. Remote Sensing of Environment. 1997;**59**(3):440-451. DOI: 10.1016/s0034-4257(96)00112-5

[27] Jordan CF. Derivation of leaf-area index from quality of light on the forest floor. Ecology. 1969;**50**(4):663-666. DOI: 10.2307/1936256

[28] Agapiou A, Hadjimitsis DG. Vegetation indices and field spectroradiometric measurements for validation of buried architectural remains: Verification under area surveyed with geophysical campaigns. Journal of Applied Remote Sensing. 2011;**5**(1):053554. DOI: 10.1117/1.3645590

[29] Sepp EM. Deeply Buried Facilities: Implications for Military Operations. Maxwell Afb Al: Air War College; 2000. DOI: 10.21236/ada425461

[30] Steven MD, Malthus TJ, Baret F, Xu H, Chopping MJ. Intercalibration of vegetation indices from different sensor systems. Remote Sensing of Environment. 2003;**88**(4):412-422. DOI: 10.1016/j.rse.2003.08.010

[31] Melillos G, Themistocleous K, Papadavid G, Agapiou A, Prodromou M, Michaelides S, et al. Importance of using field spectroscopy to support the satellite remote sensing for underground structures intended for security reasons in the eastern Mediterranean region. In: Electro-Optical Remote Sensing X. Vol. 9988. Edinburgh, United Kingdom: International Society for Optics and Photonics; 2016. p. 99880S. DOI: 10.1117/12.2240714

Chapter 6

Robust Guidance Algorithm against Hypersonic Targets

Jian Chen, Yu Han and Yuan Ren

Abstract

This chapter presents a robust guidance algorithm for intercepting hypersonic targets. Since the differential of the line-of-sight rate is more sensitive to the target maneuver, a nonlinear proportional and differential guidance law (NPDG) is given by employing the differential of the line-of-sight rate produced by a nonlinear tracking differentiator. Based on the NPDG, a fractional calculus guidance law (FCG) is presented by utilizing the differential definition of fractional order. On the basis of interceptor-target relative motions, the stability criteria of the guidance system of the FCG are deduced. In different target maneuver and noisy cases, simulation results verify that the proposed guidance laws have small miss distances and the FCG has a stronger robustness.

Keywords: hypersonic target, target maneuver, fractional order control, guidance law, stability criteria

1. Introduction

In recent years, many countries are vigorously developing hypersonic weapons in near space, such as the United States (AHW, HTV-2, X-51 and X-43), India (HSTDV and RLV-TD), China (WU-14) and Russia (GLL-31). Because of its ultra-high speed and non-fixed trajectory, the hypersonic weapon has become a great strategic threat to homeland air defense [1–5]. The hypersonic vehicle flies over 5 Mach in the near space covering distances of 20–100 km. Compared with the ballistic missile, the hypersonic weapon is usually designed in a lifting body to obtain stronger maneuverability. Traditional defense systems against cruise missiles in the atmosphere cannot reach the near space. Thereby, the near space hypersonic weapon is a threat to the current defense system.

There are mainly two kinds of hypersonic vehicles. One is the air-breathing cruise vehicle [6]. Its maneuverability is relatively weaker, thus its interception is relatively easier as its trajectory is predictable. The other is the gliding entry vehicle [7]. At the entry stage, its velocity is up to 25 Mach at maximum. In the entry phase, it is able to glide thousands of kilometers in the near space without any power. In the terminal phase, a dive attack is performed to the target on the ground [8]. Therefore, its trajectory is not predictable and its interception is a challenge. A lot of research on entry guidance techniques with no-fly zone constraints has been conducted for hypersonic weapons [9, 10]. However, there are few research works on intercepting these vehicles [11]. Consequently, new technical challenges are raised to intercept these weapons [12].

The proportional navigation guidance law (PNG) for interception has a big disadvantage of the guidance command being behind the target maneuver [13]. Actually, PNG is a proportional controller belonging to the PID controller family. Since $\ddot{q}$ embodies the target maneuver, to add a differential part $K_D \cdot \ddot{q}$ into the PNG is reasonable. Thus, the proportional and differential (PD) controller is utilized to formulate the guidance law in a hypersonic pursuit-evasion game.

By introducing fractional calculus to PID control, the fractional order PID control has become an emerging field since the 1990s [14]. Fractional calculus is a generalization of the classical integer order calculus. There are mainly three fractional calculus definitions, including Riemann-Liouville (RL) definition, Grünwald-Letnikov (GL) definition and Caputo definition. Since the Gamma function and precise solution of fractional order equations are developed, fractional calculus has appeared in the control field [15, 16]. Like integer order PID controllers, the fractional order PID controller can also be classified into PI^{λ}, PD^{μ} and $PI^{\lambda}D^{\mu}$ (λ and μ represent fractional orders). Compared to integer order PD controller concerned in this chapter, the fractional order PD^{μ} controller has the following advantages. First, a fractional order controller has greater control flexibility. There are proportional and differential fractional order μ in the PD^{μ} controller. The selection of fractional order makes it more flexible than the integer order PD controllers. Secondly, fractional order makes the controller more robust. Fractional order controller is insensitive to the parameter uncertainties of the controller and controlled plant. Even if the system parameters change a lot, a fractional order controller can still work well.

The memory function and stability characteristic make the fractional order PID controller widely applicable in the field of aircraft guidance and control [15, 16], such as pitch loop control of a vertical takeoff and landing unmanned aerial vehicle (UAV) [17], roll control of a small fixed-wing UAV [18], perturbed UAV roll control [19], hypersonic vehicle attitude control [20], aircraft pitch control [21], deployment control of a space tether system [22], position control of a one-DOF flight motion table [23], and vibration attenuation to airplane wings [24]. The viscosity of the atmosphere interacting with air vehicles has given the aircrafts the similar aerodynamics to the fractional order systems, thus the fractional order PID control theory is appropriate to design aircraft guidance and control systems.

Han et al. designed a fractional order strategy to control the pitch loop of a vertical takeoff and landing UAV. Simulations verified that the proposed controller was superior to an integer order PI controller based on the modified Ziegler-Nichols tuning rule and a general integer order PID controller in robustness and disturbance rejection [17]. Luo et al. developed a fractional order PI^{λ} controller to control the roll channel of a small fixed-wing UAV. From both simulation and real flight experiments, the fractional order controller outperformed the modified Ziegler-Nichols PI and the integer order PID controllers [18]. Seyedtabaii applied a fractional order PID controller to the roll control of a small UAV in dealing with system uncertainties, where the aerodynamic parameters are often approximated roughly [19]. Song et al. proposed a nonlinear fractional order proportion integral derivative ($NFOPI^{\lambda}D^{\mu}$) active disturbance rejection control strategy for hypersonic vehicle flight control. The proposed method was composed of a tracking-differentiator, an $NFOPI^{\lambda}D^{\mu}$ controller and an extended state observer. Simulations showed that the proposed method made the hypersonic vehicle nonlinear model track-desired commands quickly and accurately, and it has robustness against disturbances [20]. Kumar et al. developed the fractional order PID (FOPID) and integer order PID controllers using multi-objective optimization based on the bat algorithm and differential evolution technique. The proposed controllers were applied to the aircraft

DOI: http://dx.doi.org/10.5772/intechopen.84655

pitch control. Simulations demonstrated that the FOPID controller using multi-objective bat-algorithm optimization had better performance than others [21]. Ref. [22] proposed a fractional order tension control law for deployment control of a space tether system, and its stability was proved. Ref. [23] realized a fractional order controller for position control of a one-DOF flight motion table. The flight motion table was used for simulating the rotational movement of flying vehicles. Experiments showed that tracking of a position profile using fractional order controller was feasible in real time. Ref. [24] presented a tuning method of a fractional order proportional derivative controller based on three points of the Bode magnitude diagram for vibration attenuation. An aluminum beam replicating an airplane wing verified the proposed controller.

However, not much effort has been made to deal with the pursuit-evasion problem against target maneuver and guidance noise with the fractional order PID controller. Ye et al. presented a 3D extended PN guidance law for intercepting a maneuvering target based on fractional order PID control theory and demonstrated that the air-to-air missile had a smaller miss distance to a maneuvering target [25]. However, in their research, the velocity of the missile was twice as much as that of the target, and the noise impacting on the guidance state (such as line-of-sight rate) was not taken under consideration, which limits the proposed algorithm's practical engineering applications. For this reason, based on a nonlinear proportional and differential guidance law (NPDG) and fractional calculus technique, a fractional calculus guidance law (FCG) is proposed to intercept a hypersonic maneuverable target in this chapter. It is assumed that the velocity of the interceptor is same as that of the hypersonic target, which means the target can evade as fast as the interceptor, and the guidance noise of the line-of-sight rate is considered.

The rest of this chapter is organized as follows. Section 2 formulates the FCG and the system stability condition is given. Numerical experiments are carried out in Section 3, and Section 4 concludes this work.

2. Guidance law design

2.1 Definition of the NPDG

The PNG is given by

$$a_M(t) = K_P V_R(t)\dot{q}(t), \tag{1}$$

where $\dot{q}$ is the line-of-sight (LOS) angular rate, $a_M(t)$ is the normal acceleration command of the interceptor, $V_R(t)$ is the approaching velocity between the interceptor and the target, and K_P is the proportional coefficient.

For compensating the negative influence of the target maneuver, the LOS acceleration $\ddot{q}$ is considered. A nonlinear proportional and differential guidance law (NPDG) is presented as

$$a_M(t) = K_P V_R(t)\dot{q}(t) + K_D V_R(t)\ddot{q}(t), \tag{2}$$

where K_D is the differential coefficient.

A nonlinear tracking differentiator is used to estimate $\ddot{q}$. The state equation is given by

$$\begin{cases} \dot{x}_1 = x_2, \\ \dot{x}_2 = -K\,\text{sgn}\left(\dfrac{x_1 - \dot{q}_m + |x_2|x_2}{2K}\right), \end{cases} \tag{3}$$

where K is the estimation coefficient, $\dot{q}_m(t)$ is the LOS rate measured by the seeker, $\dot{q}_m(t)$ and $\ddot{q}_m(t)$ are estimated by x_1 and x_2, namely $x_1 = \hat{\dot{q}}_m(t)$ and $x_2 = \hat{\ddot{q}}_m(t)$. It is not easy to determine the value of K. If K is larger, the estimation will be more precise and the phase lag will be less, but the estimation will be noisier. Therefore, a fractional calculus guidance law is presented.

2.2 Formulation of the FCG

The Grünwald-Letnikov (GL) fractional differential definition to formulate the FCG is presented as

$$ {}_a^G D_t^\mu f(t) = \lim_{h \to 0} \frac{1}{h^\mu} \sum_{k=0}^{\frac{t-a}{h}} (-1)^k \frac{\Gamma(\mu+1)}{k!\Gamma(\mu-k+1)} f(t-kh), \tag{4} $$

which extends it from integer order to fractional order.

On dividing the continuous interval $[a, t]$ of $f(t)$ with step $h = 1$, and setting $n \in \{1, 2, ..., t\text{-}a\}$, the difference equation of the fractional differential signal of $f(t)$ is given by

$$ \frac{d^\mu f(t)}{dt^\mu} \approx f(t) + (-\mu) f(t-1) + \frac{(-\mu)(-\mu+1)}{2} f(t-2) + \ldots + \frac{\Gamma(-\mu+1)}{n!\Gamma(-\mu+n+1)} f(t-n). \tag{5} $$

According to definitions of the NPDG and GL, the FCG is proposed as

$$ a_M(t) = K_P V_R(t) x(t) + K_D V_R(t) \frac{d^\mu x(t)}{dt^\mu}, \tag{6} $$

where μ is the fractional order, $\frac{d^\mu x(t)}{dt^\mu}$ is the fractional differential of $x(t)$, and $x(t) = \dot{q}(t)$.

In the FCG, the future state of the GL fractional differential of $\dot{q}$ depends on the previous and current states. But in the NPDG, the future state $\ddot{q}$ only depends on the current state. It indicates that the fractional order part is a filter with the "memory" characteristic. The FCG runs like a filter, which is insensitive to the noises, and shows robustness to disturbances.

2.3 Stability criteria

As shown in **Figure 1**, the target and interceptor are located in the same plane, XOY, where M and T denote the interceptor and target; θ_M and θ_T represent flight

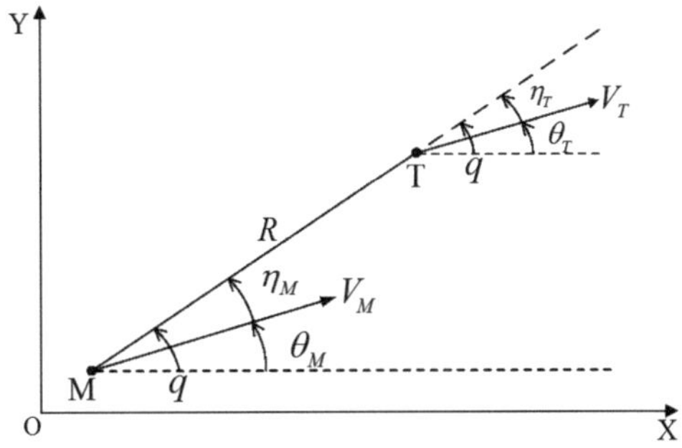

Figure 1.
Planar endgame engagement geometry.

path angles of the interceptor and target; η_M and η_T represent their heading angles; V_M and V_T represent their velocities; R represents the relative distance between them; and q is the line-of-sight angle of the interceptor.

The relative motion equations are given by

$$\dot{q} = \frac{1}{R}(V_M \sin\eta_M - V_T \sin\eta_T), \tag{7}$$

$$V_R = \dot{R} = -V_M \cos\eta_M + V_T \cos\eta_T, \tag{8}$$

$$q = \theta_M + \eta_M = \theta_T + \eta_T. \tag{9}$$

Differentiating Eq. (7), and substituting Eq. (8) and Eq. (9) into it, we have

$$R\ddot{q} + 2\dot{R}\dot{q} = \dot{V}_M \sin(\dot{q} - \dot{\theta}_M) - \dot{V}_T \sin(\dot{q} - \dot{\theta}_T) + \dot{\theta}_T V_T \cos(\dot{q} - \dot{\theta}_T) - \dot{\theta}_M V_M \cos(\dot{q} - \dot{\theta}_M). \tag{10}$$

2.3.1 Linearization

For a nonlinear problem Eq. (10), classic stability analysis theories such as the Routh-Hurwitz stability criterion for linear systems cannot be applied directly. Linearization must be done first.

Considering the practical situation, the values of $\dot{V}_M$, $\dot{V}_T$ and $\dot{\theta}_T$ will approach zero in the endgame [26]. Then, the nonlinear system Eq. (10) can be simplified into a linear system:

$$R\ddot{q} + 2\dot{R}\dot{q} \approx -\dot{\theta}_M V_M \cos(q - \theta_M). \tag{11}$$

From Eq. (11), the transfer function of the guidance system is obtained as

$$\frac{\dot{q}(s)}{\dot{\theta}_M(s)} = \frac{-V_M \cos(q - \theta_M)}{Rs + 2\dot{R}} = \frac{-K_G}{T_G - 1}, \tag{12}$$

where

$$K_G = \frac{V_M \cos(q - \theta_M)}{2|\dot{R}|}, \; T_G = \frac{R}{2|\dot{R}|}.$$

Thus, we get

$$\dot{q}(s) = \frac{-K_G}{T_G s - 1}\dot{\theta}_M(s). \tag{13}$$

From Eq. (6), since $a_M = V_M\dot{\theta}_M$, we have

$$\dot{\theta}_M = \frac{V_R}{V_M}(K_P\dot{q} + K_D\dot{q}s^{\mu}). \tag{14}$$

Substituting Eq. (14) into Eq. (13), the characteristic equation of the fractional calculus guidance system becomes

$$\frac{V_R}{V_M}K_G K_D s^{\mu} + T_G s + \left(\frac{V_R}{V_M}K_G K_P - 1\right) = 0. \tag{15}$$

2.3.2 Stability analysis

In stability analysis of Eq. (15), the Hurwitz stability criterion is appropriate to be employed.

Lemma 1: Hurwitz stability criterion [27]

For an nth-degree polynomial characteristic equation:

$$D(s) = a_0 s^n + a_1 s^{n-1} + \cdots + a_{n-1} s + a_n = 0 \ (a_0 > 0) \tag{16}$$

the necessary and sufficient stability condition, of system (16), is

$$\Delta_1 = a_1 > 0,\ \Delta_2 = \begin{vmatrix} a_1 & a_3 \\ a_0 & a_2 \end{vmatrix} > 0,\ \Delta_3 = \begin{vmatrix} a_1 & a_3 & a_5 \\ a_0 & a_2 & a_4 \\ 0 & a_1 & a_3 \end{vmatrix} > 0,\ \cdots,\ \Delta_n > 0. \tag{17}$$

That is, the order of principal minor determinants and the main determinant of the system (16) is positive.

Thus, based on the Hurwitz stability criterion, the necessary and sufficient stability condition of system (15) becomes

$$a_0 = \frac{V_R}{V_M} K_G K_D > 0, \tag{18}$$

$$\Delta_1 = a_1 = T_G > 0, \tag{19}$$

$$\Delta_2 = \begin{vmatrix} a_1 & a_3 \\ a_0 & a_2 \end{vmatrix} = \begin{vmatrix} T_G & 0 \\ \frac{V_R}{V_M} K_G K_D & \frac{V_R}{V_M} K_G K_P - 1 \end{vmatrix} = T_G \times \left(\frac{V_R}{V_M} K_G K_P - 1 \right) > 0. \tag{20}$$

That is

$$\begin{cases} \frac{V_R}{V_M} K_G K_D > 0, \\ T_G > 0, \\ \frac{V_R}{V_M} K_G K_P - 1 > 0. \end{cases} \tag{21}$$

Since $K_P > 0$ and $K_D > 0$, K_P can be preset as 4. As a consequence, we have $\cos(q\text{-}\theta_M) > 0.5$, that is $\cos\eta_M > 0.5$. It concludes

Theorem 1: When the interceptor's heading angle η_M is in the range of $-60°$ to $+60°$, the fractional calculus guidance system remains stable.

3. Numerical simulations

3.1 Simulations design

For intercepting a hypersonic weapon, a space-based surveillance satellite and a ground-based X band radar or a marine X band radar should detect the target as early as possible to provide the interceptor enough time to launch from the ground or the aerial carrier. In the terminal phase of a hypersonic weapon, its velocity is too high to be intercepted. For example, the speed of a gliding entry vehicle is up to 25 Mach at maximum during a dive attack to the ground target. Thus, the interception is usually designed in the gliding or cruising phase in the near space of a hypersonic weapon before its terminal phase (i.e., before a dive attack happens); then, the interceptor-target initial position and encounter condition is designed to be a head-to-head encounter. In the gliding or cruising phase in the near space of a hypersonic weapon, its velocity is relatively low (about 5 Mach), and its maneuvering amplitude cannot exceed 5 g due to the reduced aerodynamic efficiency since the atmosphere is thin in the near space, but the time instant that the hypersonic weapon starts maneuvering is flexible and adjustable for evading the interceptor's pursuit. Our preliminary studies and experiments show that it is not good for the hypersonic weapon to start maneuvering as early as possible during a pursuit-evasion game, and it is better for the hypersonic weapon to start maneuvering when the interceptor is close to it in the endgame. For the maneuvering mode of the hypersonic weapon to evade the interceptor's pursuit, the step maneuver and square maneuver are preferred to the ramp maneuver and sine maneuver since they can provide the hypersonic weapon the maximum evading acceleration instantly.

Based on the above analysis, the simulation parameters for a hypersonic pursuit-evasion game are set as: the interceptor-target initial position and heading condition is planned in a head-to-head engagement, and the initial relative distance R = 30,000 m, V_T = 5 Mach, which is along the negative X-axis; V_M = 5 Mach, and its initial direction is aimed at the target, that is $\theta_M = q$; the initial LOS angle q is 10°; the interceptor's maximum normal acceleration is 15 g; μ is set to the best value of 0.5 based on experience. Obviously, $\eta_M = q - \theta_M = 0° \in [-60°, 60°]$. The fractional calculus guidance system is stable based on Theorem 1.

According to authentic maneuvering characteristics of a hypersonic weapon in the gliding or cruising phase in the near space when the interceptor is close to it, its maneuver equations are given by.

Case 1: Step maneuver

$$a_T = 5\text{g}, \quad t \geq 8\text{s}, \tag{22}$$

Case 2: Square maneuver

$$a_T = \begin{cases} 5\text{g}, & t \in [2k + 6, 2k + 7)\text{s}, \\ -5\text{g}, & t \in [2k + 7, 2k + 8)\text{s}, \end{cases} \tag{23}$$

where a_T is the norm acceleration of the target, t is the time index and $k \in$ N. The target maneuvers are shown in **Figures 2** and **3**.

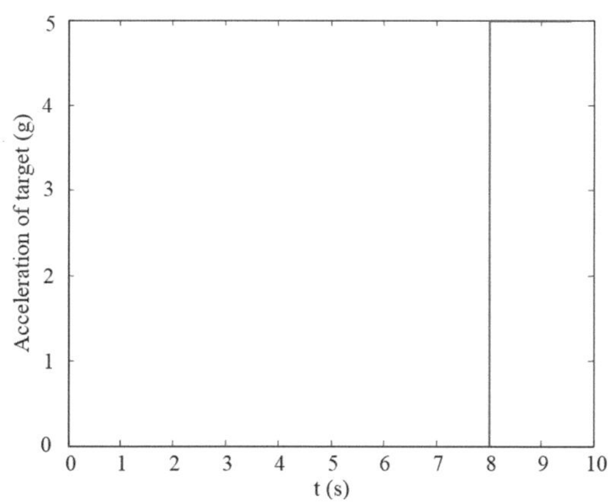

Figure 2.
Step maneuver of the target (case 1).

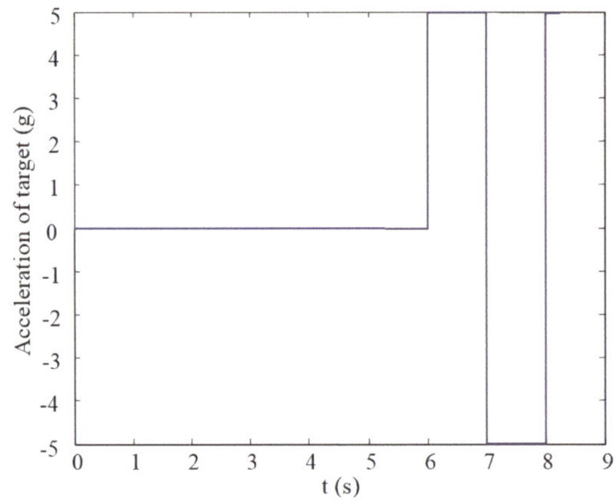

Figure 3.
Square maneuver of the target (case 2).

3.2 Interception accuracy

The trajectories, line-of-sight rates and guidance commands of the interceptor and target are shown in **Figures 4–9**. From **Figures 4** and **5**, since the velocities of the interceptor and target are hypersonic (5 Mach), the amplitude of the target maneuvers is 5 g which cannot change the velocities and trajectories of the target a lot in a limited endgame time. Thus, there is no big difference between the trajectories of the target between **Figures 4** and **5**. From **Figures 6** and **7**, the line-of-sight rates constrained by the FCG are much smaller than those constrained by the NPDG. And the line-of-sight rates of the NPDG are always non-convergent. From **Figures 8** and **9**, the guidance commands of the FCG are much smoother than those of the NPDG, which are more appropriate for the interceptor's autopilot to track. The reason is that the NPDG uses a nonlinear tracking differentiator Eq. (3) to estimate $\ddot{q}$. In Eq. (3), K is the coefficient of the estimator. The larger the K is, the more precise the estimation is and the less the phase lag is, but the noisier the

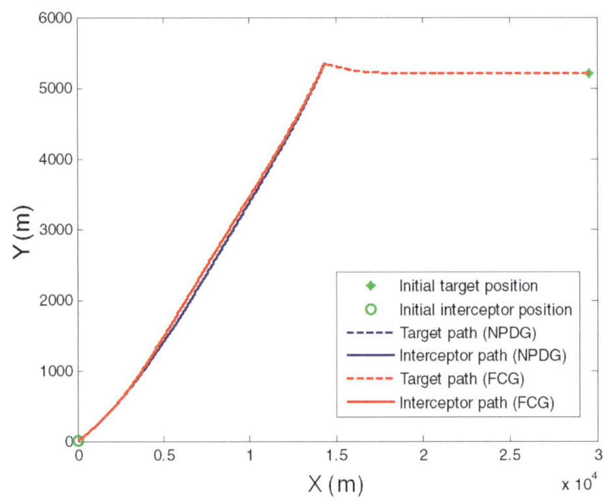

Figure 4.
Trajectories of the interceptor and target (case 1).

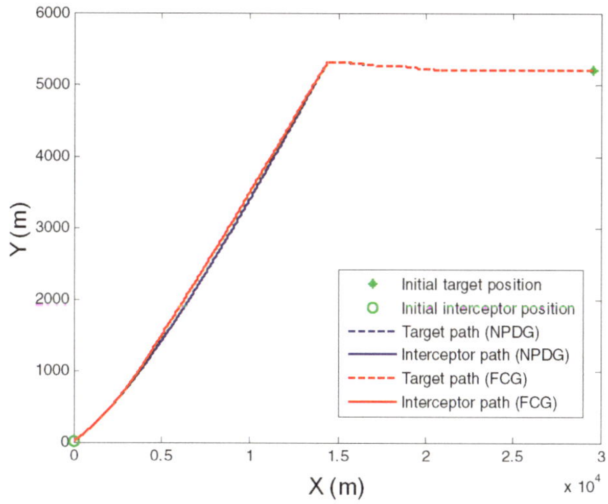

Figure 5.
Trajectories of the interceptor and target (case 2).

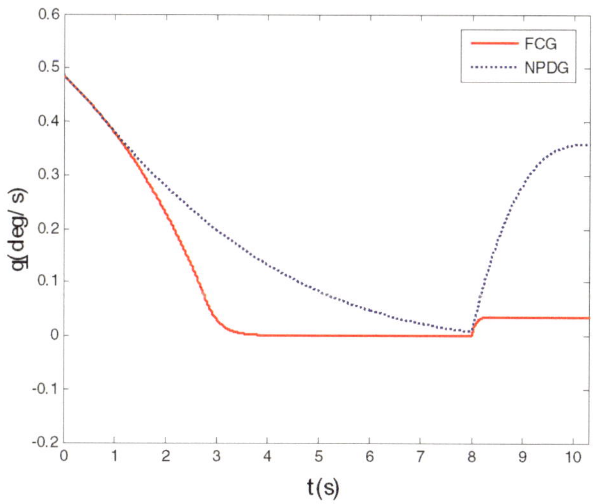

Figure 6.
Line-of-sight rates (case 1).

estimation is. Comparing **Figure 9** with **Figure 8**, the guidance command of the NPDG in case 2 is noisier than that in case 1, which means the target maneuver of case 2 is more challenging to the NPDG than that of case 1. It is also validated by the results in **Table 1** that the miss distance of the NPDG in case 2 is larger than that of the NPDG in case 1. However, the target maneuver of case 2 has little influence on the interception accuracy of the FCG, since the miss distance of the FCG in case 2 is even smaller than that of the FCG in case 1, which indicates the superiority of the FCG.

Numerical results are demonstrated in **Table 1**. The FCG has the minimum miss distance under different scenarios. In case 1, the miss distance of the FCG is 0.0322 m, which is 91% less than that of the NPDG (0.3406 m). In case 2, the miss distance of the FCG is 0.0294 m, which is 93% less than that of the NPDG (0.4151 m).

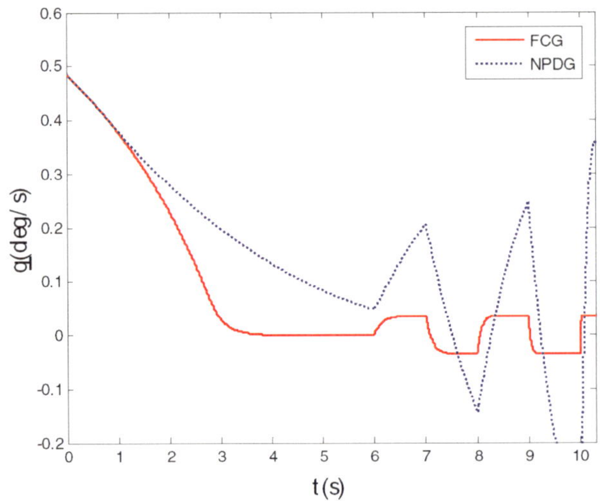

Figure 7.
Line-of-sight rates (case 2).

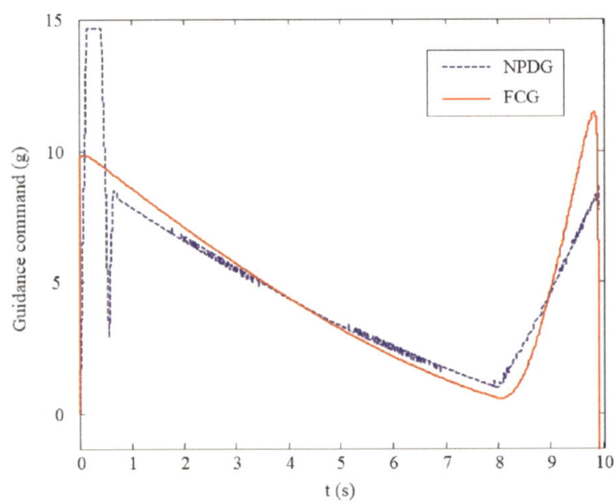

Figure 8.
Guidance commands (case 1).

3.3 Stability

In case 1, when pre-setting the simulation parameters, if the initial flight path angle θ_M is set as 40°, 70° and 75°, and other parameters remain unchanged, obviously, the heading angle $\eta_M = q - \theta_M$, will be −30°, −60° and −65°, respectively. The stabilities of the fractional calculus guidance system with the FCG can be analyzed based on Theorem 1.

As shown in **Figures 10–12**, when the heading angle ηM belongs to the closed interval [−60°, 60°], the interceptor can hit and kill the target; when the heading angle ηM is beyond the closed interval [−60°, 60°], the interception mission fails.

Simulation results are compared and summarized in **Table 2**. The miss distances increase as the heading angle goes beyond the closed interval [−60°, 60°]; when the heading angle η_M is −60°, it is a critical condition. The experimental results in **Table 2** validate the conclusion of Theorem 1.

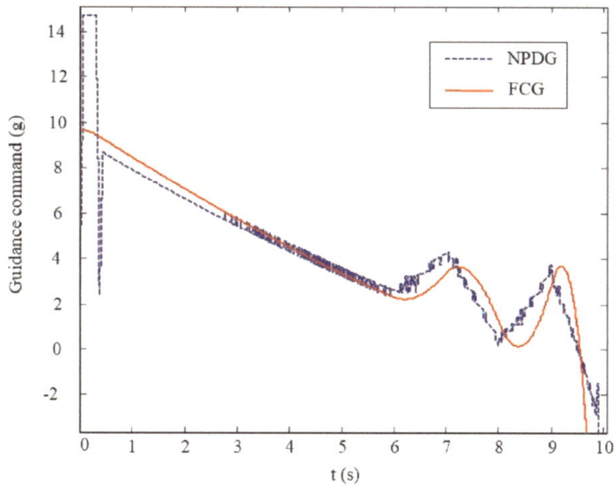

Figure 9.
Guidance commands (case 2).

Guidance law	Case 1: miss distance (m)	Case 2: miss distance (m)
NPDG	0.3406	0.4151
FCG	0.0322	0.0294

Table 1.
Performance evaluation of guidance laws.

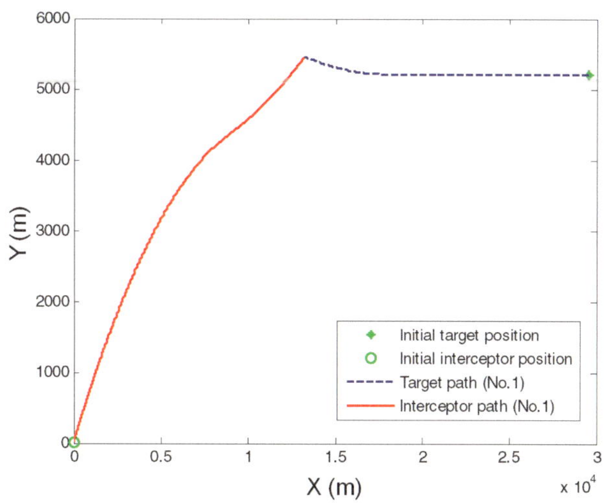

Figure 10.
Trajectories of the interceptor and target (No. 1 $\eta_M = -30°$).

3.4 Robustness

In case 1, three white noises are added into $\dot{q}$ to run 50 groups of the Monte Carlo simulations, including the amplitudes of 0.5°/s, 1.5°/s and 2.5°/s. The total number of tests is 50. The miss distance distributions of the NPDG and the FCG with a noise of 0.5°/s, 1.5°/s and 2.5°/s are shown in **Figures 13–18**.

From **Figures 13, 15** and **17**, it can be seen that the miss distances of the NPDG obviously increase as the noise increases. Similarly, from **Figures 14, 16** and **18**, the

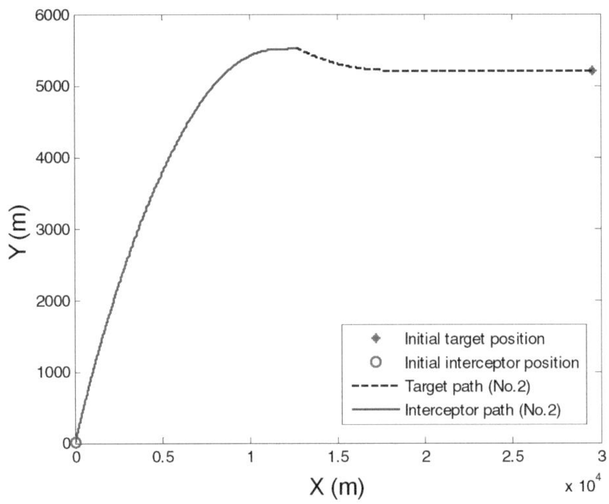

Figure 11.
Trajectories of the interceptor and target (No. 2 $\eta_M = -60°$).

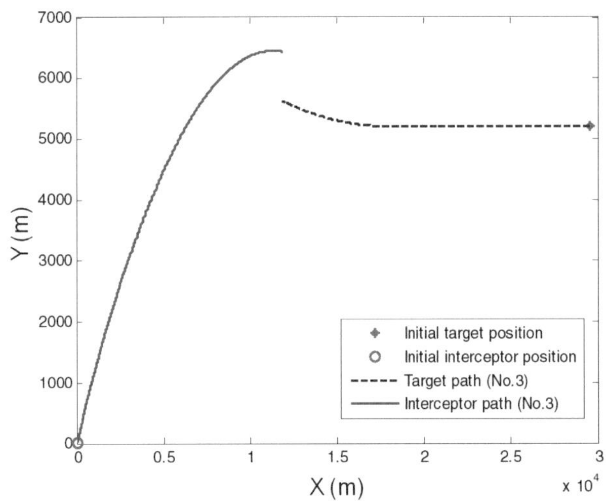

Figure 12.
Trajectories of the interceptor and target (No. 3 $\eta_M = -65°$).

No.	Heading angle η_M (°)	Stability	Miss distance (m)
1	−30	Stable	0.1060
2	−60	Stable	8.9125
3	−65	Unstable	820.7977

Table 2.
Stability analysis.

miss distances of the FCG slightly increase as the noise increases. These phenomena indicate the effect of noise impacting on the miss distances of both the NPDG and the FCG. Moreover, comparing **Figure 14** with **Figure 13**, comparing **Figure 16** with **Figure 15**, and comparing **Figure 18** with **Figure 17**, the miss distances of the

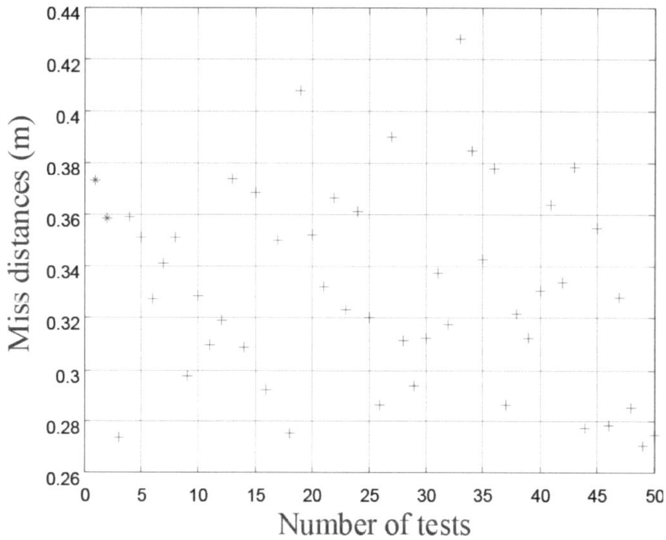

Figure 13.
Miss distance distribution of the NPDG with a noise of 0.5°/s.

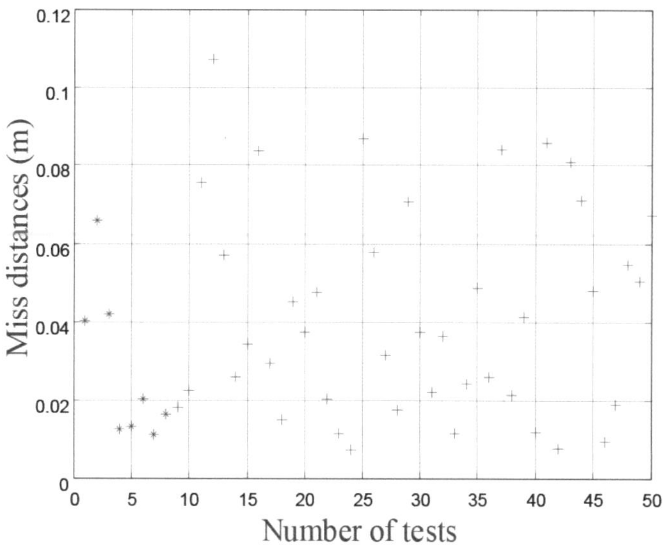

Figure 14.
Miss distance distribution of the FCG with a noise of 0.5°/s.

FCG are always smaller than those of the NPDG, which indicates the stronger robustness of the FCG.

Statistical results are indicated in **Table 3**. Obviously, compared with the NPDG, the FCG has a better robustness to the guidance noises.

To summarize the interception accuracy and robustness experiments, a conclusion can be drawn. The unique filtering properties of the fractional calculus guidance law make its interception accuracy and robustness better. For intercepting a hypersonic weapon, introducing the differential signal of the line-of-sight rate as the guidance information can effectively suppress the target maneuvers, and it has a good robustness, which can make it a feasible guidance strategy. The specifications are as follows:

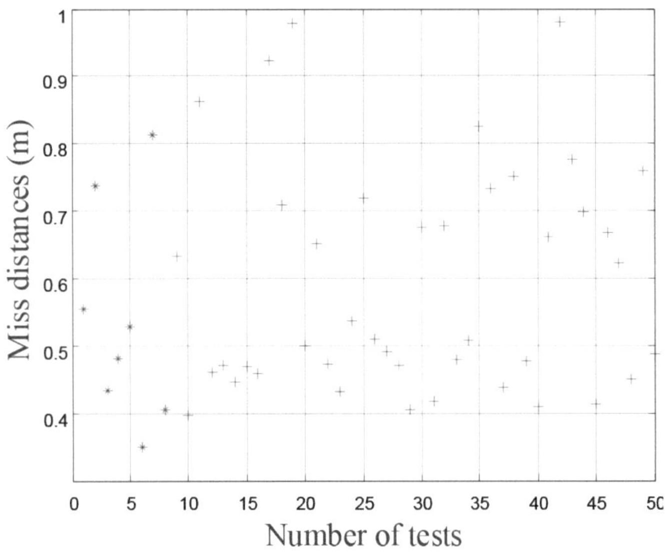

Figure 15.
Miss distance distribution of the NPDG with a noise of 1.5°/s.

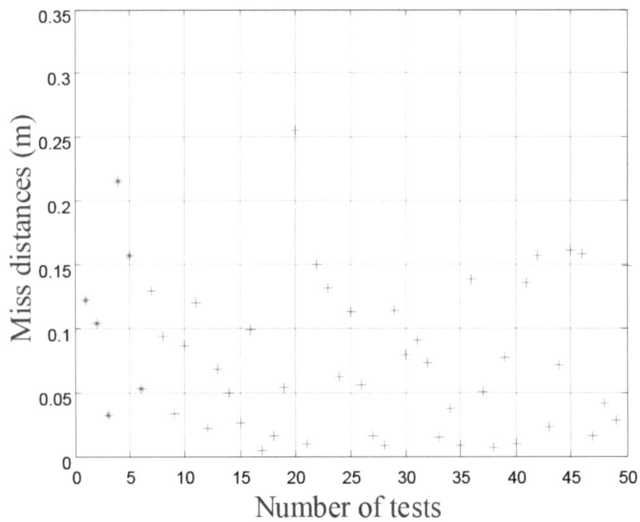

Figure 16.
Miss distance distribution of the FCG with a noise of 1.5°/s.

1. The FCG can improve the guidance accuracy. Compared with the NPDG, it has a better feasibility, since the NPDG requires the measurement of $\ddot{q}$, while this angular acceleration usually cannot be directly measured by the interceptor's seeker.

2. The robustness of the FCG is better than that of the NPDG. The FCG using the fractional differential of $\dot{q}$ improves the precision of the estimation. The filtering capability of the fractional order part in the FCG provides good stability to the system in a hypersonic pursuit-evasion game under noisy conditions.

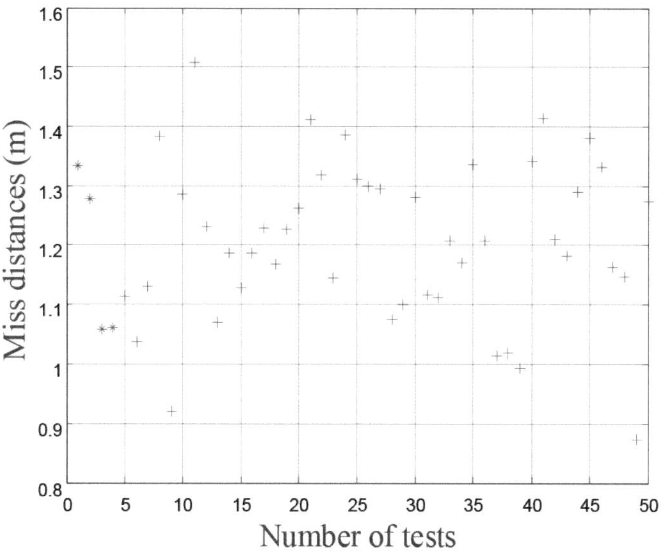

Figure 17.
Miss distance distribution of the NPDG with a noise of 2.5°/s.

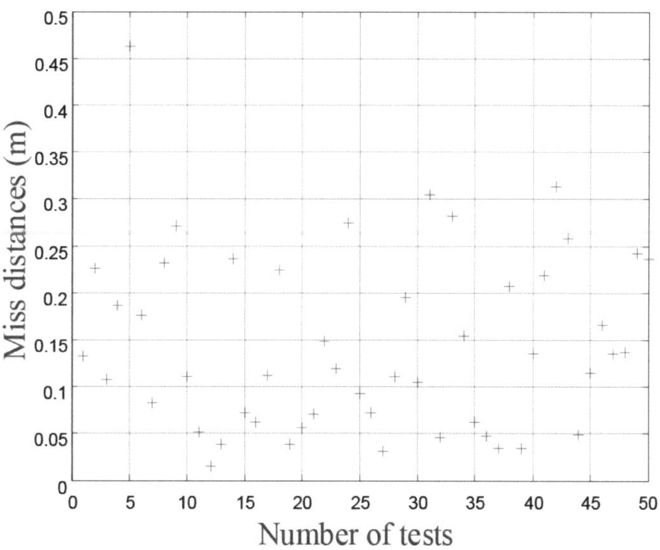

Figure 18.
Miss distance distribution of the FCG with a noise of 2.5°/s.

Noise (°/s)	Guidance law	Expectation (m)	Variance (m)
0.5	FCG	0.0396	6.7768e−004
0.5	NPDG	0.3322	0.0014
1.5	FCG	0.0786	0.0036
1.5	NPDG	0.5842	0.0274
2.5	FCG	0.1457	0.0091
2.5	NPDG	1.0092	0.2044

Table 3.
Statistical results of the miss distances under noisy conditions.

4. Conclusions

This chapter first discusses how to solve the problem of intercepting the hypersonic maneuvering target without greatly increasing the complexity degree of the guidance system. Based on the axiom that the response to the target maneuver of the differential signal of the line-of-sight rate is faster than that of the line-of-sight rate, a nonlinear proportional and differential guidance law is designed using the differential derivative of the line-of-sight rate. Based on the differential definition of fractional calculus, a fractional calculus guidance law is designed on the basis of the NPDG. In the simulation experiments of interception accuracy and robustness, both the NPDG and the FCG demonstrate guaranteed guidance performances. The influence of noises impacting on the guidance system is studied. Both of the guidance laws can effectively intercept hypersonic maneuvering targets while reducing the impact of noise signals. Furthermore, the method obtaining the fractional differential signal of $\dot{q}$ in the FCG is better than the method estimating the $\ddot{q}$ in the NPDG.

In conclusion, under the premise of not greatly increasing the complexity degree of the guidance system, introducing the differential signal of the line-of-sight rate to formulate the novel guidance laws can help meet the precision needed to intercept a hypersonic weapon. The FCG is superior to the NPDG in interception accuracy and robustness to guidance noises.

Acknowledgements

This work is supported by National Key R&D Program of China (Grant Nos. 2016YFC0400207, 2017YFD0701003 from 2017YFD0701000, and 2016YFD0200702 from 2016YFD0200700), the Jilin Province Key R&D Plan Project 2017YFD0701000, and 2016YFD0200702 from 2016YFD0200700), the Jilin (Grant Nos. 20180201036SF and 20170204008SF), and the Chinese Universities Scientific Fund (Grant Nos. 10710301, 1071-31051012, 1071-31051361, and 2019TC108).

Conflict of interest

The authors declare no conflict of interest.

Author details

Jian Chen[1,2], Yu Han[3*] and Yuan Ren[4]

1 College of Engineering, China Agricultural University, Beijing, China

2 Beijing Key Laboratory of Optimized Design for Modern Agricultural Equipment, Beijing, China

3 College of Water Resources and Civil Engineering, China Agricultural University, Beijing, China

4 Space Engineering University, Beijing, China

*Address all correspondence to: yhan@cau.edu.cn

References

[1] Romaniuk S, Grice F. The Future of US Warfare. New York: Taylor & Francis; 2017

[2] Besser H, Huggins M, Zimper D, Goge D. Hypersonic vehicles: State-of-the-art and potential game changers for future warfare. In: Brochure of 2016 NATO Science & Technology Symposium; March 2016; Brussels. Paris: NATO Science & Technology Organization; 2016. DOI: 10.13140/RG.2.1.3360.8566

[3] Majumdar S. BrahMos: A hypersonic future? Vayu Aerospace and Defence Review. 2017;**1**:116

[4] Butt Y. A hypersonic nuclear war is coming. New Perspectives Quarterly. 2016;**33**(1):51-54

[5] Bartles C. Russian threat perception and the ballistic missile defense system. Journal of Slavic Military Studies. 2017;**30**(2):152-169

[6] Murugan T, De S, Thiagarajan V. Validation of three-dimensional simulation of flow through hypersonic air-breathing engine. Defence Science Journal. 2015;**65**(4):272-278

[7] Li G, Zhang H, Tang G. Maneuver characteristics analysis for hypersonic glide vehicles. Aerospace Science and Technology. 2015;**43**:321-328

[8] Xie D, Wang Z, Zhang W. A new strategy of guidance command generation for re-entry vehicle. Defence Science Journal. 2013;**63**(1):93-100

[9] Wang T, Zhang H, Tang G. Predictor-corrector entry guidance with waypoint and no-fly zone constraints. Acta Astronautica. 2017;**138**:10-18

[10] He R, Liu L, Tang G, Bao W. Rapid generation of entry trajectory with multiple no-fly zone constraints. Advances in Space Research. 2017;**60**(7):1430-1442

[11] Liu C, Liu C, Tuan P. Algorithm of impact point prediction for intercepting reentry vehicles. Defence Science Journal. 2016;**56**(2):129-146

[12] Vathsal S, Sarkar A. Current trends in tactical missile guidance. Defence Science Journal. 2005;**55**(2):265-280

[13] Tyan F. Analysis of general ideal proportional navigation guidance laws. Asian Journal of Control. 2016;**18**(3):899-919

[14] Das S. Functional Fractional Calculus. Berlin: Springer; 2011

[15] Chen Y, Vinagre B, Xue D. Fractional-Order Systems and Controls: Fundamentals and Applications. London: Springer; 2010

[16] Li Z, Liu L, Dehghan S, Chen Y, Xue D. A review and evaluation of numerical tools for fractional calculus and fractional order controls. International Journal of Control. 2017;**90**(6):1165-1181

[17] Han J, Di L, Coopmans C, Chen Y. Pitch loop control of a VTOL UAV using fractional order controller. Journal of Intelligent & Robotic Systems. 2014;**73**(1–4):187-195

[18] Luo Y, Chao H, Di L, Chen Y. Lateral directional fractional order $(PI)^{\alpha}$ control of a small fixed-wing unmanned aerial vehicles: Controller designs and flight tests. IET Control Theory and Applications. 2011;**5**:2156-2167

[19] Seyedtabaii S. New flat phase margin fractional order PID design: Perturbed UAV roll control study. Robotics and Autonomous Systems. 2017;**96**:58-64

[20] Song J, Wang L, Cai G, Qi X. Nonlinear fractional order proportion integral derivative active disturbance rejection control method design for hypersonic vehicle attitude control. Acta Astronautica. 2015;**111**:160-169

[21] Kumar P, Narayan S, Raheja J. Optimal design of robust fractional order PID for the flight control system. International Journal of Computer Applications. 2015;**128**(14):31-35

[22] Sun G, Zhu Z. Fractional-order tension control law for deployment of space tether system. Journal of Guidance, Control, and Dynamics. 2014;**37**(6):2057-2061

[23] Zarei M, Arvan M, Vali A, Behazin F. Design and implementation of fractional order controller for a one DOF flight motion table. In: Proceedings of the 36th Chinese Control Conference; 26–28 July 2017; Dalian. New York: IEEE; 2017. pp. 11395-11400

[24] Birs I, Folea S, Muresan C. An optimal fractional order controller for vibration attenuation. In: Proceedings of the 25th Mediterranean Conference on Control and Automation; 3–6 July 2017; Valletta. New York: IEEE; 2017. pp. 2473-3504

[25] Ye J, Lei H, Li J. Novel fractional order calculus extended PN for maneuvering targets. International Journal of Aerospace Engineering. 2017: 5931967-1-5931967-9

[26] Ben-Asher J, Speyer J. Games in aerospace: Homing missile guidance. In: Basar T, Zaccour G, editors. Handbook of Dynamic Game Theory. Cham: Springer; 2017. pp. 1-28. DOI: 10.1007/978-3-319-27335-8_25-1

[27] Dorf R, Bishop R. Modern Control Systems. Boston: Pearson; 2011